ROUTLEDGE LIBRARY EDITIONS: GEOLOGY

Volume 2

AEOLIAN GEOMORPHOLOGY

AEOLIAN GEOMORPHOLOGY
Binghamton Geomorphology Symposium 17

Edited by
WILLIAM G. NICKLING

Routledge
Taylor & Francis Group

LONDON AND NEW YORK

First published in 1986 by Allen & Unwin Inc.

This edition first published in 2020
by Routledge
2 Park Square, Milton Park, Abingdon, Oxon OX14 4RN

and by Routledge
52 Vanderbilt Avenue, New York, NY 10017

Routledge is an imprint of the Taylor & Francis Group, an informa business

British Library Cataloguing in Publication Data
A catalogue record for this book is available from the British Library

ISBN: 978-0-367-18559-6 (Set)
ISBN: 978-0-429-19681-2 (Set) (ebk)
ISBN: 978-0-367-21054-0 (Volume 2) (hbk)
ISBN: 978-0-367-21055-7 (Volume 2) (pbk)
ISBN: 978-0-429-26515-0 (Volume 2) (ebk)

Publisher's Note
The publisher has gone to great lengths to ensure the quality of this reprint but points out that some imperfections in the original copies may be apparent.

Disclaimer
The publisher has made every effort to trace copyright holders and would welcome correspondence from those they have been unable to trace.

Aeolian Geomorphology

Edited by
William G. Nickling

Proceedings of the 17th Annual Binghamton
Geomorphology Symposium, September 1986

Boston
ALLEN & UNWIN
London   Sydney

This volume was prepared, proofed and passed for press by the Editors and Contributors

Allen & Unwin Inc.,
8 Winchester Place, Winchester, Mass. 01890, USA

Allen & Unwin (Publishers) Ltd,
40 Museum Street, London WC1A 1LU, UK

Allen & Unwin (Publishers) Ltd,
Park Lane, Hemel Hempstead, Herts HP2 4TE, UK

Allen & Unwin (Australia) Ltd,
8 Napier Street, North Sydney, NSW 2060, Australia

First published in 1986

Library of Congress Cataloging-in-Publication Data

"Binghamton" Geomorphology Symposium (17th : State University of New York at Binghamton)
    Aeolian geomorphology.
("Binghamton" symposia in geomorphology ; no. 17)
Bibliography: p.
Includes index.
1. Aeolian processes--Congresses.
    2. Geomorphology--Congresses. I. Nickling,W. II. Title. III. Series.
GB611.B56 1986   551.3'7   86-17492
ISBN 0-04-551133-0 (alk. paper)

British Library Cataloguing in Publication Data

Binghamton Geomorphology Symposium
(17th: 1986)
    Aeolian geomorphology.—(The 'Binghamton' symposia in geomorphology International series, ISSN 0261-3174; no. 17)
1. Wind erosion
I. Title II. Nickling, William III. Series
551.3'7   QE597
ISBN 0-04-551133-0

Printed in Great Britain by
Butler & Tanner Ltd, Frome and London

Preface

Over four decades have passed since R.A. Bagnold first published his classic work *The Physics of Windblown Sand and Desert Dunes*. This outstanding pioneer work, which set the foundation for almost all subsequent aeolian research, was complemented by the volumous writings of W.S. Chepil during the 1940's and 1950's (summarized in Chepil and Woodruff 1963). Chepil's applied research provided both theoretical and empirical observations on the primary variables that influence the erosion of agricultural soils by wind.

Despite this excellent foundation, aeolian research decreased significantly during the 1960's with few major contributions being made to the literature. The work of Belly (1964) and Williams (1964) on the mechanics of sand transport by wind are noteable exceptions.

It is apparent however, that the early 1970's brought renewed interest to the investigation of aeolian processes and landforms. Although this interest resulted from many factors, two appear to be of particular significance. The increased awareness of the desertification problem, particularly in Africa and Asia, necessitated a clearer understanding of aeolian processes so that appropriate control measures could be developed and implemented. A second, but unrelated factor, was the detailed information gained from the Mariner and Viking spacecraft missions that showed the importance of aeolian processes in the development of the Martian landscape. Subsequent detailed investigations into aeolian processes and landforms on other planets (e.g. Breed *et al.* 1979, Greeley *et al.* 1980, 1982, Iversen and White 1982) through theoretical modelling and simulation, has provided new insight and understanding of similar processes on Earth.

The renewed interest and major advancements in the study of aeolian processes and landforms resulted in this topic being selected as the theme for the 1986 Binghamton Geomorphology Symposium. Although an attempt was made to show the range of current interests in aeolian phenomena it is by no means a comprehensive accounting of the diverse research currently under investigation. The following papers do however, clearly indicate that the study of aeolian processes and landforms is not constrained by discipline boundaries and is of interest to geologists, physical geographers, soil scientists, meteorologists and engineers.

The sixteen papers of this volume can be grouped into six major themes. The first four papers focus on the mechanics of sediment transport and indicate the importance of wind tunnel modelling in aeolian research. Brian Willetts and Ann Rice, on the basis of wind tunnel tests, show that particle shape has a pronounced effect on the saltation path and the nature of the intervening bed collisions. Their results suggest that bed collision is much less effective in sustaining saltation for platy shaped grains than for more compact quartz grains.

D. Gillette and P. Stockton describe new instrumentation to directly measure mass and momentum fluxes of particles transported in wind tunnels. They observed in their wind tunnel tests that mass and momentum fluxes changed markedly with distance from the leading edge of the sand deposit. This variation was shown to be related to the change in friction velocity resulting from the interaction between the fluid flow and the saltating particles. They also suggest that the variation in mass and momentum fluxes is affected by the finite ceiling height of the wind tunnel which limits momentum transfer from above. Scale effects in wind tunnel modelling are also discussed by J. Iversen. In his paper, Iversen demonstrates the usefulness of dimensionless parameters in the investigation of saltation physics and modelling of drift geometry at small scales. He suggests that although dimensionless deposition or deflation rate may be a function of several dimensionless parameters, the Froude number is one of the most important.

Bruce White discusses a series of experiments carried out in a specially designed high pressure wind tunnel that dynamically simulates particle motion on the surface of Venus. Comparison of experimental particle saltation trajectories with theoretical calculations show very good agreement, giving support to the theoretical saltation model he has developed for Venusian conditions.

The second group of papers focus on the morphology and development of desert dunes. Haim Tsoar and Jens Møller describe the role of vegetation in the formation and maintenance of linear sand dunes. They suggest that the destruction of the vegetation cover on linear dunes results in their transformation to linear-braided or sief forms. Ian Livingstone, on the basis field observations, proposes a qualitative model of linear dune dynamics. He argues that linear dune form is the response to the pattern of wind speed change and does not necessarily require secondary flow phenomenum as other models suggest. Helga Besler describes the complex development of a longitudinal dune which has recently been cut across by the construction of the Toshka Canal.

Dune development in coastal environments are discussed in the two following papers which form the third major theme. Aeolian processes and dune characteristics on the developed shoreline of Westhampton Beach, New York are examined by K. Nordstrom, J.

McCluskey and P. Rosen. In their paper they describe the effect
of coastal processes and control structures on the development and
morphology of dunes in this area. The following paper   by A.
Orme and V. Tchakerian outlines the development of Quaternary
dunes along the Pacific coast of California. In their study area the
authors identify four distinct phases of dune development during
the Quaternary although the dune sequences have similar
orientations and internal structures, differences in dune morphology
vegetation cover and grain size characteristics allow for the
differentiation of individual dune phases.

The fourth group of papers focuses on landforms and bedforms
associated with aeolian erosion and deposition. Gary Kocurek, in
his analysis of bedforms, points out that a significant proportion of
the aeolian rock record consists of low angle strata (≤ 15 degrees)
which are distinct from higher angle cross strata usually associated
with true dune forms. He argues that these bedforms represent
deposits in interdune areas, sand sheets and aprons. he does point
out however, that dunes can also produce low-angle strata when
aprons or lower plinth deposits are preserved. Low-angle deposits
consisting entirely of wind-ripple laminae are best developed on
oblique crescentic, linear and star dunes were secondary flow
produces considerable along slope wind.

Despite our greatly improved understanding of aeolian phenomena,
few studies have assessed the effect of topography on aeolian
processes and associated landform development. Ron Greeley, on
the basis of extensive wind tunnel simulations detailed field studies
and theoretical consideration investigates erosional and depositional
features associated with three distinct landforms: domical hills,
raised-rim craters and yardangs. For each of these landforms
consideration is given to the nature of the air flow around the
obstruction and the resultant erosional/depositional forms. His
observations suggest that for the case of domical hills and
raised-rim craters a horseshoe vortex wraps around the obstruction
forming distinctive zones of potential erosion and deposition.

The paper by C. McKenna-Neuman and R. Gilbert is significant in
that it is one of very few studies that investigates active aeolian
processes in a periglacial environment. Their field observations
suggest that aeolian sediment transport on arctic proglacial sandurs
is most significant during the fall and early winter months when
aerial exposure of the sediments and wind velocities are greatest.
During the fall, sediment transport is strongly controlled by short
term variation in surface moisture content. In contrast, the supply
of sediment for winter transport appears to be provided by two
mechanisms: (1) grain-by-grain release of the particles from the
frozen sandur surface as result of the sublimation of interstitial ice
and (2) the abrasive action of grains saltating across the frozen
surface.

The fifth theme presented in the collected papers centres on the
generation of major dust storms. N. Middleton, A. Goudie and G.

Wells through the combined analysis of standard meteorological observations, remotely sensed data and deep sea cores, have identified major locations of dust storm activity on a global scale. In addition, their data provide a general indication of regional dust storm frequency.   However, they point out that dust storm frequency and sediment flux from a given region may vary considerably from one year to the next because of minor climatic shifts and changes in landuse patterns.

The problem of temporal and spatial variability of dust storm activity on a more localized scale is addressed by A. Brazel, W. Nickling and J. Lee in their paper on the effects of antecedent moisture conditions on dust storm generation at Phoenix, Arizona. Although several authors have shown a strong correlation between mean annual precipitation and mean annual dust storm frequency at the regional/global scales no significant correlation was found between these variables for single station data collected at Phoenix during the period 1948-1984.   Their data however, did indicate a significant relationship between winter precipitation and mean annual dust frequency.   They argue that high evaporation rates in summer tend to off-set short term, moderately wet periods that may be antecedent to dust events.   Moreover, evidence of vegetation response to increased winter precipitation suggests that interannual variability of vegetation cover is a significant factor in dust storm generation at this localized scale.

The final two papers by Richard Marston and Ronald Dorn represent the more applied aspects of aeolian research.   Richard Marston reports on the effect of training excercises using heavy tanks and ground-based artillery on accelerating wind erosion at Fort Bliss Military Reservation in south central New Mexico.   The heavy tanks truncate and bisect dunes fixed by mesquite (*Prosopis glandulosa*) and enhance interdune deflation by mechanical break-up of the natural surface crust composed of carbonate salts and fine grained particles.   His observations indicate that the effect of maneuver traffic on threshold wind velocity are relatively minor and short-lived.   Precipitation events subsequent to vehicular disturbances re-establish the surface crust causing and increase threshold velocity required to initiate particle motion.   He does point out however, that the effects of maneuvers on dune microtopography are more persistent due to the preservation of impact scars by subsequent surface crusting.

Ronald Dorn suggests that micromorphology of rock varnish is controlled largely by the abundance of aeolian dust and that superimposition of dissimilar micromorhphologies reflects fluctuations in the aeolian environment. Varnishes were sampled from over 100 arid, semi-arid and humid sites in western North America and analysed using scanning electron microscopy (SEM), energy-dispersive analysis of X-rays (EDAX) and particle induced X-ray emission (PIXE).   In addition, Dorn has used cation-ratio dating of rock varnishes formed on ventifacts to provide a minimum age for the cessation of aeolian abrasion. His analysis of varnishes on K-Ar dated rocks from eastern California suggest that this area experienced three lengthy dusty periods alternating with two long periods of less abundant dust during the last 200 000 BP.

William G. Nickling
Department of Geography
University of Guelph

References

Bagnold, R.A. 1941. *The physics of blown sand and desert dunes*, London: Methuen.

Breed, C.S., M.J. Grolier and J.F. McCauley 1979. Morphology and distribution of common 'sand' dunes on Mars. Comparison with the Earth. *Journal of Geophysical Research* 84, 8183-204.

Belly, P.Y. 1964. *Sand Movement by Wind*. US Army Corps of Engineers, Coastal Engineering Research Centre, Technical Memoir No. 1.

Chepil, W.S. and N.P. Woodruff 1963. The physics of wind erosion and its control. *Advances in Agronomy*, 15, 211-302.

Greeley, R., R. Leach, B.R. White, J.D. Iversen and J.B. Pollack 1980. Threshold windspeeds for sand on Mars: Wind tunnel simulations *Geophysical Research Letters* 7, 121-4.

Greeley, R., R. Leach, S.H. Williams, B.R. White, J.B. Pollack, D.H. Krinsley and J.R. Marshall 1982. Rates of wind abrasion on Mars. *Journal of Geophysical Research*, 87, 10009-24.

Iversen, J.D. and B.R. White 1982. Saltation threshold on Earth, Mars, and Venus. *Sedimentology* 29, 111-19.

Williams, G. 1964. Some aspects of the eolian saltation load. *Sedimentology* 3, 257-87.

Acknowledgements

I would like to acknowledge the financial support for the Symposium provided by the National Science and Engineering Research Council of Canada. My appreciation is also extended to M.R. Moss, Chairman, Department of Geography, University of Guelph for providing logistical and financial support during the preparation of this manuscript.

In particular I would like to thank: Heidi Facklemann for her extensive assistance with the technical editing, Becky Morrison for typing the entire manuscript, Marie Puddister for her excellent work in preparing the layout for the camera ready copy, Peter McCaskell who assisted in the preparation of the type-set copy and Chris Scime for his assistance with the cartographic work. Their efforts are greatly appreciated.

My thanks are also extended to J. Gillies, D. Maenpaa, J. Alberico and C. Finlan for their assistance in various aspects of the Symposium organization. The support and advice of Athol Abrahams, Lorraine Oak and J. Vitek on the planning of the Symposium and preparation of the proceedings is gratefully acknowledged. Finally I would like to thank each of the sixteen authors for their excellent papers.

William G. Nickling

Contents

Preface
Acknowledgements

xvi

1

Collision in aeolian transport: the saltation/creep link

Brian B. Willetts and M. Ann Rice

Abstract

Collision, it is contended, is the mechanism which is necessary to investigate in order to better understand the launch conditions for saltation, creep transport and ripple formation, and surface re-sorting.

Observations of landing and take-off velocities at individual collisions are incorporated in approximate calculations of momentum change for three size fractions of each of two sands. One sand is a quartz sand with compact grain shape, and the other, a shell sand with more platy grains, particularly in larger fractions. Calculations are made for horizontal bed and for collision with a bed having a local adverse slope.

Grain shape is found to have a marked effect on collision. The shell sand grains lose rather less forward momentum on average than do quartz grains, and undergo a much smaller change of vertical momentum. It is inferred that collision is much less effective in sustaining a saltating load in the case of shell sand than in that of quartz sand.

Adverse bed slope reduces the loss of forward momentum at collision, but increases more dramatically the change of vertical momentum. Therefore, as one moves up a ripple back towards the crest, creep activity increases, but the vigour of launch of new saltations diminishes. These saltations will land downwind with uneven spatial distribution. The high rate of creep activity near the crest will produce a coarser than average grain population there.

Introduction

It is well known that grains of wind-blown sand move by three processes called, respectively, suspension, saltation, and creep. Saltation is believed to be particularly important. Grains in saltation receive momentum from the wind and deliver it to the bed at a rapid rate. This keeps the bed surface in a disturbed state, prompting further saltation and stimulating both creep and the release of fine material into suspension. Measurement of the transport rate of suspended material is virtually impossible. Of the remaining mobile material, roughly three-quarters of the transport rate is by saltation and roughly one quarter by creep (Bagnold 1941, Willetts & Rice in press a).

Its acknowledged' importance has led to several studies of saltation. Films and multiple exposure still photographs have been taken which enabled detailed measurements to be made on the higher portion of the trajectory (White & Schultz 1977, Hunt & Nalpanis in press), and produced data from which velocity and acceleration have been computed. Several theoretical models of the saltation trajectory have been published (e.g. White & Schultz 1977, Hunt & Nalpanis in press, Jensen & Sorensen 1983). There are two significant uncertainties in using such models. One concerns the initial conditions for the trajectory (velocity and angle of launch), and the other concerns the rate and effect of particle spin. Both of these features are closely related to the outcome of the collision which usually precedes the saltation.

Creep has been studied less than saltation. It takes place at the base of a layer of moving sand which becomes deep and populous when transport is active. In these circumstances it is difficult to make detailed observations. Creep transport rates can be obtained by using traps designed to exclude most saltating grains, and grain velocities estimated by using such traps in conjunction with dyed grains originating at a known distance upwind (Willetts & Rice in press a). However, the amount of quantitative information available about creep is very small. Yet a large proportion of the moving grains in unit bed area are moving in creep, and the process must, therefore, be influential in ripple formation, in the re-sorting of the surface population, and in the release of dust. It is generally accepted that creep motions are instigated by collision, as saltating grains return to the bed.

Collision then, is at the root both of the unknown features of saltation, and of the important, but obscure process of creep. Nevertheless, there have been a limited number of studies of collision. Tsuchiya and Kawata (1972) incorporated intelligent estimates of the outcome of collision in a calculation of repeated saltation. Rumpel (1985) performed a simplified analysis of the mechanics of idealized collisions in order to generate data by means of which he could demonstrate the construction and use of probability distribution matrices for the outcome of saltation sequences. Willetts & Rice (in press b) made direct observations of individual collisions and produced sets (of restricted size) of measured data which can be used in the way in which Rumpel used his artificial data.

It is the purpose of this paper to explore the links between saltation, collision, and creep.   Data derived from direct observation of collision will be used, and will concern the behaviour of two sands of different characteristic grain shape. (One is a quartz sand and the other a shell-derived sand with a large proportion of platy grains). The difficulties of observation limit the quality and quantity of the data, and therefore, the security of the conclusions which can be drawn.   However, some manipulation of the data at this stage is instructive in revealing the potential benefit obtainable by the pursuit of further data of this kind.

Experimental method

Collisions were filmed at approximately 3000 frames per second with the camera positioned to give a virtually true view of the plane of saltation of the incident grains.   These grains were artificially introduced into the wind tunnel in such a way that they saltated once or twice on the fixed, roughened floor of the tunnel before colliding with the surface of a tray of sand.   It was this last collision which was filmed in each case.   The rate and manner of introduction of sand was controlled, so that the number of grains seen in each photographic frame was sufficiently small for each to be identifiable, and capable of being followed in a sequence of frames.   Thus, the stream direction and vertical components of grain displacement in a short time interval could be obtained, and thence the magnitude and direction of the grain velocity component in the vertical plane of flow.   The experimental set-up has been described in more detail elsewhere (Willetts & Rice in press b).

Such velocity vector information was calculated for the incident grain and for visible grain movements consequent upon the collision.   Figure 1 identifies, by suffix, the approach velocity v_1 and angle α_1, a velocity v_3 and angle α_3 of richochet (or rebound), and ejection velocities of previously stationary grains v_n and α_n.   The symbol n may take values from 4 to 10, depending on the number of ejections caused by the collision under observation.   Approximately 100 collisions were observed for each experimental condition.

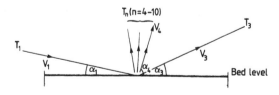

Figure 1 Schematic view of collision. Impact of the incident grain at v_1 and a_1 gives rise to several dislodgements at v_n and a_n. The incident grain ricoshets at v_3 and a_3.

The sand tray was filled with one of two parent dune sands. The introduced grains which saltated onto it were drawn in any one experiment from one of three fractions sieved out from the appropriate parent population. Figure 2 illustrates the size distribution of each of the sands used, and identifies the fractions which are called "coarse", "medium" and "fine" in the remainder of the paper.

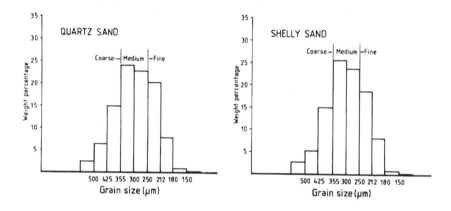

Figure 2 Grain size distributions of the parent sands A-quartz and B-shell.

The two sands differed from one another primarily in grain shape. This was assessed for each sand by measuring for approximately 100 grains, the longest dimension, a, the smallest orthogonal dimension, c, and the dimension, b, normal to both a and c. Shape differences are shown in Figure 3 in two ways: in triangular graphs of c/a vs a-b/a-c, and by means of average values of the minimum projection sphericity, Ψ_p (Sneed & Folk 1958). The latter gives a single figure for each population by which they can be compared; the former illustrates the range of individual grain shapes present in each sand. Sand A can be seen to contain predominantly compact grains, whereas sand B has a much larger proportion of platy grains.

The difference between the relative densities of the sands was too small to be influential. Coefficient of restitution was not measured; it may be different in the two cases, and might then affect inter-grain collisions. However, the impulses generated in collision between grains of such small mass are very small, and therefore, unrecovered distortion of the material seems unlikely. Therefore, differences in coefficient of restitution are thought unlikely to have significant effect.

Because of the artificial element in the experimental arrangement, corroboration was sought that the approach vector to the impact adequately replicated that found in natural sand driving. The observed approach angles fell within the range 9°-13°, which has been shown to conform well with the findings of other observers (Willetts & Rice in press b).   Since this was so, and since our attention is confined to grain movements very near the point of impact (to the exclusion of other features of the saltation trajectory), the data reported is believed to be undistorted by the artificiality of the set-up.

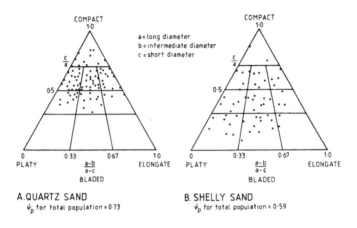

Figure 3 Shape characteristics of the parent sands: form diagram after Sneed and Folk (1958).

Results

Figure 4 shows for the medium fraction of each sand, the distribution of take off velocities separately for ricocheting grains, and for freshly disturbed grains at a wind speed corresponding to $u_* = 39$ cm/s.   The sample is rather small in the case of the Achmelvich sand. Take-off velocities of both kinds are smaller for the platier shell sand than for the quartz sand. This is true of all size fractions, as can be seen in Table 1.

The angles of descent, α_1, and of ascent (α_3 to α_{10}) are all slightly smaller on averge for the shell sand than for the quartz sand. In the case of both sands, the ricochet angle α_3 has mean and modal values much lower than the conventionally accepted take-off angle of 90°, and this is true to a smaller extent of the take-off angles of dislodged grains α_4 - α_{10} (see Fig. 5 & Table 1). Table 1 also shows that the quartz grains on average dislodge more stationary grains at each collision than do the shell grains.

Since the angle and magnitude of the velocity is known for each grain near collision, the components parallel and normal to the bed can be determined for individual grains. This has been done, and

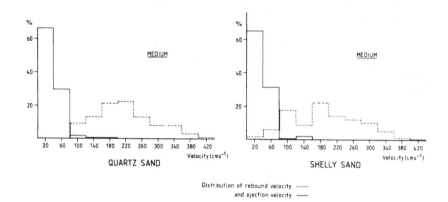

Figure 4 Distributions of v_3 and v_n, the initial velocities of ricochet and dislodgement.

Table 1 A comparative summary of the mean values obtained for the quartz (Q) and shelly (S) sands.

		Coarse	Medium	Fine
$v_1 \, cm s^{-1}$	Q	360.0	356.0	394.0
	S	356.0	345.0	384.0
$V_3 \, cm s^{-1}$	Q	225.0	216.0	240.0
	S	220.0	192.0	211.0
$V_n \, cm s^{-1}$	Q	40.0	36.0	31.0
	S	46.0	33.0	40.0
$a_1 {}^\circ$	Q	12.7	11.7	9.6
	S	11.4	11.2	9.2
$a_3 {}^\circ$	Q	31.3	24.9	33.4
	S	16.3	19.3	20.7
$a_n {}^\circ$	Q	52.0	54.0	54.0
	S	48	52.0	44.0
% of grains causing dislodgement	Q	82.0	69.0	51.0
	S	72.0	59.0	34.0
No. ejections per impact	Q	3.1	1.5	0.9
	S	21.1	1.2	0.5

the mean values for each sand type and incident grain size are shown in Table 2. The terms $\Sigma v_n \sin \alpha_n$ and $\Sigma v_n \cos \alpha_n$ contain, for each collision, the sum of the appropriate component for the grains dislodged at that collision up to an arbitrary limit of seven. It is very rare for there to be more than seven dislodgements at a collision, so a negligible amount of data are excluded by the limit.

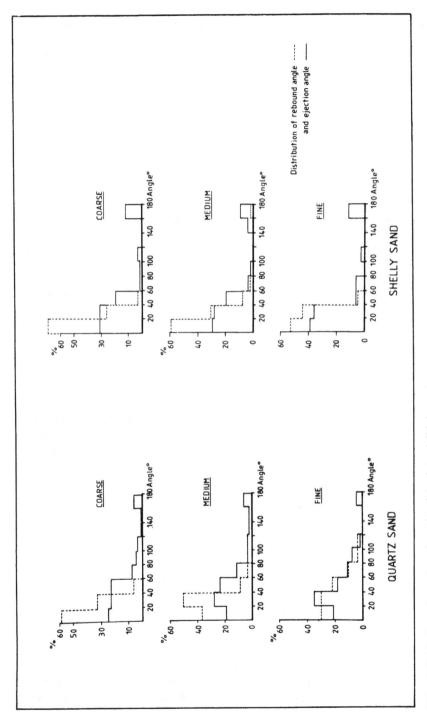

Figure 5 Distributions of α_3 and α_n, the angles of ricochet and dislodgement.

Table 2 Mean forward and vertical components of velocity for movements at collisions originated by incident grains in each size fraction. Units are cms^{-1}.

Sand	Incident Grain Fraction	$v_1 \sin a_1$	$v_3 \sin a_3$	$v_n \sin a_n$	$v_1 \cos a_1$	$v_3 \cos a_3$	$v_n \cos a_n$
A (quartz)	Coars	77.3	71.6	63.8	350.0	208.2	75.6
	Medium	70.4	80.7	31.5	347.7	194.1	31.4
	Fine	62.1	106.3	15.7	387.5	198.2	15.7
B(shell)	Coarse	64.2	55.9	37.9	348.8	197.9	60.2
	Medium	60.6	55.2	16.6	338.3	175.9	22.2
	Fine	58.8	72.6	12.1	378.6	192.1	21.2

Momentum changes at collision are of considerable interest because of their links with forces exerted on the bed. While they cannot be rigorously calculated because the size of freshly disturbed grains is not known, reasonable estimates of momentum loss can be made using a mean value of mass for each grain class. Table 3 lists mean grain mass for each fraction of each sand and for the parent population of each sand.

Discussion

The experiments have concerned individual collisions, and compilations of the resulting observations contain interesting statistical information about the consequences of collision involving grains of different size and shape. Such data have been used elsewhere in discussion of the velocity vectors of approach and departure (Willetts & Rice in press b, c) with reference primarily to the dynamics of saltation. Each grain leaving the bed, whether by ricochet or by initial dislodgement, undertakes a saltation unless it becomes suspended. Therefore the range of departure velocities is of distinct relevance to the study of saltation. In particular, the distributions of values of the ratios:

$$v_3 \sin 3 : v_1 \sin 1 \text{ and } v_n \sin n : v_1 \sin 1$$

have been used to ascertain probabilities, in various circumstances, of growth of saltation amplitude after collision. For both sands at $u_* = 39$ cm/s, the coarse fraction gave rise to saltation of increasing height in less than half the observed collisions (43% for sand A, 45% for sand B). In the case of both medium and coarse fractions, grains of the platy sand (B) were significantly less likely to undertake consecutive saltations of increasing height than were grains of the quartz sand (A).

In the circumstances of this experiment, fine grain saltations

Table 3 Mean grain mass for each fraction of each sand. The figures in columns 2 and 5 were obtained by weighing a known number of grains; those in columns 4 and 7 are derived from them.

Fraction	Sieve aperture	Quartz Sand (A)			Shell Sand (B)		
		Grain Mass $g \times 10^5$	Proportion present by weight (%)	Mean Grain Mass $g \times 10^5$	Grain Mass $g \times 10^5$	Proportion present by weight (%)	Mean Grain Mass $g \times 10^5$
Coarse	> 500	30.36	2.5		14.56	2.5	
	425–500	17.86	6.0	12.58	9.93	5.0	8.34
	355–425	10.70	15.5		7.42	15.0	
Medium	300–355	5.86	24.0		4.33	26.0	
	250–300	3.35	22.5	4.30	2.81	24.0	3.44
Fine	212–250	2.23	20.0		2.30	18.5	
	180–212	1.50	8.0	1.90	1.24	8.0	1.75
	150–180	1.21	1.5		0.83	1.0	

For all grains $\bar{m}$ = 3.54 x 10^{-5} g

For all grains $\bar{m}$ = 3.04 x 10^{-5}

were growing in height on average (more vigorously in the case of sand A) while coarse grain saltations were diminishing. For medium sand, the quartz grain saltation was increasing slightly in vigour, while the shell sand saltation was decaying. The proportions of ricochets which give rise to saltation growth are summarized in Table 4 for the three size fractions of each sand. Since the dislodgement rate per collision is also lower for shell sand (Table 1), it is clear that inter-grain collision is a much less effective mechanism for sustaining the saltating population in this case than it is for quartz sand. The difference is attributable to shape differences between the sands.

Table 4 The proportion of collisions in which the vertical components of take-off velocity is greater than the vertical component of incident volocity.

	Coarse	Medium	Fine
Sand A (quartz)	0.43	0.52	0.71
Sand B (shell)	0.45	0.46	0.59

Discussion thus far has concerned grain velocities, which can be unambiguously derived from the photographic records. The effect of the colliding grains on the bed is better examined in terms of momentum changes. Because the mass of each individual grain is not known, this cannot be done with the degree of confidence attributable to velocity comparisons. However, a number of approximate, but useful results can be obtained by employing the mean values of grain mass given in Table 3.

Consider the collision of a coarse grain which approaches the bed at v_1 and α_1, as defined in Figure 1, and leaves it after ricochet at v_3 and α_3. Grains of unknown mass are dislodged with initial velocity v_n and angle α_n. Suppose that the distribution of velocity of initial disturbance is independent of the size of dislodged grains. The momentum changes of the grains which are seen to move can be calculated approximately as follows:

$$\text{Loss of forward momentum} = m_c (v_1\cos\alpha_1 - v_3\cos\alpha_3) - \Sigma\bar{m}v_n\cos\alpha_n \quad (1)$$

$$\text{Change of vertical momentum} = m_c (v_1\sin\alpha_1 + v_3\sin\alpha_3) + \Sigma\bar{m}v_n\sin\alpha_n \quad (2)$$

where: m_c is the mean mass for coarse fraction grains
$\bar{m}$ is the mean mass for all grains.

Corresponding equations can be written for the medium and fine fractions, using in each case the mean mass for the appropriate fraction.

The first two rows of figures in Table 5 show the mean value of each of these momentum changes for each fraction of each sand.

The third row shows the mean ratio of the forward : vertical momentum change in each case. While the forward momentum loss is significantly less for the shell than for the quartz sand in all fractions, the vertical momentum change shows a much more pronounced difference. In consequence, the ratio loss of forward momentum : change of vertical momentum, is approximately 0.75 for quartz sand and approximately 1.0 for shell sand. Expressed in terms of a dynamic friction angle, these correspond to 37° and 45°.

Table 5 Momentum changes at the bed (rows 1-3) and normalized bed shear stress (rows 4 and 5). Assumption (a) is that all grains strike the bed, the rate of striking for each fraction is proportional to the weight of that fraction in the parent population. Assumption (b) is that this is true only for the coarse and medium fractions, the fine fraction being "lost" into suspension.

Flat Bed	Sand A (quartz)			Sand B (shell)		
	Coarse	Medium	Fine	Coarse	Medium	Fine
1. Loss of forward momentum, m_f, per collision $(g.cm.s^{-1} \times 10^6)$	151.70	54.90	30.40	106.90	49.10	26.20
2. Change of momentum, m_n, normal to bed per collision $(g.cm.s^{-1} \times 10^6)$	209.90	76.20	37.60	111.60	44.80	26.70
3. $m_f \div m_n$	0.72	0.72	0.81	0.96	1.10	0.98
4. Saltation-generated bed shear stress normalized by the quartz grain value under assumption (a)						
Assumption (a) (see text)		1.00			1.02	
Assumption (b)		0.65			0.74	
5. Saltation generated normal stress normalized by quartz grain shear stress.						
Assumption (a)		1.33			1.01	
Assumption (b)		0.90			0.71	

At this point, the natural direction for the analysis to take is towards the calculation of the normal and shear stresses exerted on the bed by the saltating grain population. Unfortunately, too little is known about the number of grains in saltation and the rate at which unit area of bed is struck by them for the stresses to be calculated absolutely. However, it is possible, using information about the two sands, and some simplifying assumptions, to compare the stresses which each generates on the bed. In the last two rows of Table 5, shear and normal stress values are compared for

sands A and B on the assumptions: (a) that the rates of bed
collision for each fraction are proportional to the weight of that
fraction present in the parent sand; and (b) that this is true for
the coarse and medium fractions only, the fine fraction being
entirely lost to suspension.   The calculation is made for a given
mass of parent sand.   In the larger sieve sizes, the shell grains
have smaller average mass than the quartz ones because of their
shape, so that shell sand would generate more collisions per unit
bed area and unit time than quartz sand for the same mass
saltation load, provided that mean path lengths were the same.
This last condition is rather implausible because the subjective
impression is that shell grain trajectories tend to be longer than
quartz ones.   Nevertheless, it is retained for the purpose of these
comparisons because there is no objective data on which to frame
an alternative.   This should be borne in mind when considering the
normalized stresses listed in Table 5.

Discussion so far has concerned collision with a horizontal flat
bed, so that angles α_1 in Figure 1 can be read as inclinations to
the horizontal or to the plane of the bed.   In natural conditions,
the bed is very rarely flat, but is rippled.   Maintenance of the
ripple shape as it advances requires that transport rate increases
from the trough to the crest on the upwind face.   This face is
curved, and therefore at each point on it, the surface is presented
at a different angle to the horizontal and to a grain approaching at
a given angle to the horizontal.   It is therefore of interest to
consider the response of grain velocity and momentum changes to
variation in the angle of approach to the bed.

Suppose that the struck surface in Figure 1 is allowed to tilt
through an angle θ, so that the grain approaching at α_1 to the
horizontal now makes an angle of incidence of $(\alpha_1 + \theta)$ with the bed.
On the evidence available at present (Willetts & Rice in press b),
increase in the angle of incidence beyond 12° has little influence
on the statistics of the velocities of grains leaving the bed.   For
present purposes it will be assumed that there is no influence, i.e.
a collision at $(\alpha_1 + \theta)$ to the bed gives rise to the same distributions
with respect to the bed of v_3, α_3, v_n and α_n as would a collision
at incidence angle α_1.   On this assumption, expressions can be
written as follows for the loss of momentum parallel to the bed per
collision, and for the change of vertical momentum per collision.
The symbols m_c and $\overline{m}$ are defined as in Equations 1 and 2, and
the equations concern the collision of an incident coarse grain.

Loss of forward momentum
per collision
$$= m_c v_1 \cos(\alpha_1 + \theta) - m_c v_3 \cos\alpha_3 - \Sigma \overline{m} v_n \cos\alpha_n \qquad (3)$$

Change of vertical
momentum per collision
$$= m_c v_1 \sin\alpha_1 + m_c v_3 \sin(\alpha_3 + \theta) + \Sigma \overline{m} v_n \sin(\alpha_n + \theta) \qquad (4)$$

As in the case of Equations 1 and 2, corresponding equations can be written for the medium and fine fractions.

Changes of momentum per grain, averaged for each fraction, are shown in Table 6 for each of four bed slopes, θ. Also listed there in parenthesis, are values for each non-zero bed slope normalized with respect to the horizontal bed value for the corresponding fraction.

Table 6 Forward and vertical momentum changes per grain averaged for each class, for horizontal bed and three adverse bed gradients. In each case the figure is parenthesis is normalized with respect to the corresponding horizontal bed case.

Sand fraction	Bed Angle $\theta°$	Momentum changes (normalized in parenthesis)			
		Forward momentum		Vertical momentum	
		g cms^{-1} x 10^5		g cms^{-1} x 10^5	
Quartz coarse	0	1517	(1.00)	2099	(1.00)
	5	1415	(0.93)	2347	(1.12)
	10	1281	(0.84)	2583	(1.23)
	15	1115	(0.74	2808	(1.34)
Quartz medium	0	549	(1.00)	762	(1.00)
	5	517	(0.94)	842	(1.10)
	10	474	(0.86)	919	(1.21)
	15	420	(0.77)	991	(1.30)
Quartz fine	0	304	(1.00)	376	(1.00)
	5	290	(0.95)	412	(1.10)
	0	272	(0.89)	446	(1.19)
	15	248	(0.82)	479	(1.27)
Shell coarse	0	1076	(1.00)	1116	(1.00)
	5	1018	(0.95)	1275	(1.14)
	10	938	(0.87)	1426	(1.28)
	15	838	(0.78)	1572	(1.41)
Shell medium	0	492	(1.00)	448	(1.00)
	5	469	(0.95)	506	(1.13)
	10	438	(0.89)	561	(1.25)
	15	398	(0.81)	614	(1.37)
Shell fine	0	263	(1.00)	267	(1.00)
	5	251	(0.95)	301	(1.13)
	10	235	(0.89)	334	(1.25)
	15	213	(0.81	365	(1.37)

It can be seen that the loss of forward momentum per collision declines in each class with increase in θ, while the vertical momentum change increases with increase in θ. The sensitivity to bed slope is greater for coarse grains than for medium or fine ones in both sands. The decline of forward momentum loss is steeper for quartz than for shell sand, but the increase of the vertical momentum change with angle is more pronounced in the shell sand figures. Were it possible to multiply these momentum changes per collision by the number of collisions per second on an appropriate bed area, one could arrive at estimates of:

(a) the forward stress exerted by moving grains on the bed, and
(b) the weight of saltating sand which is supported directly by intermittent contact with the bed.

However, the collision rate is not known, and therefore stress cannot be calculated. The best that can be done is to calculate a "stress coefficient" for each fraction and each bed slope which shows how stress changes with the angle of presentation of the bed. The first step in doing this is represented by the bracketed figures in Table 6 which have already been discussed. The stress coefficient, however, must also take into account the increase in the number of collisions per unit bed area when the angle $(\alpha_1+\theta)$ increases. When compared with the collision rate per unit bed area on the horizontal bed, that on a bed inclined at θ is increased by a factor $\cos\alpha_1 \div \cos(\alpha_1+\theta)$ for the same approach rate of grains at α_1 to the horizontal. When this adjustment is made, stress coefficients are arrived at as shown in Table 7. It can be seen that the effect of the concentration of collisions as the bed angle changes offsets partially the change with bed angle of the loss of forward momentum. On the other hand, it accentuates the change with θ of vertical momentum change. With these qualifications, the effects of bed slope are similar to those noted earlier on the raw momentum changes per collision.

Table 7 Stress coefficients showing the variation with bed angle θ of the saltation produced surface shear stress (column 3) and the vertical component of stress.

Sand fraction	Bed angle $\theta°$	Stress coeficients (see text for definition)	
		Shear stress	Vertical stress component
Quartz coarse	0	1.00	1.00
	5	0.96	1.15
	10	0.89	1.30
	15	0.81	1.47
Quartz medium	0	1.00	1.00
	5	0.96	1.12
	10	0.91	1.27
	15	0 84	1.42
Quartz fine	0	1.00	1.00
	5	0.97	1.11
	10	0.94	1.24
	15	0.88	1.38
Shell medium	0	1.00	1.00
	5	0.97	1.15
	10	0.94	1.32
	15	0.89	1.51

The grain-generated boundary shear stress, or traction, is greater for horizontal beds than for adverse bed gradients. This can be interpreted in the context of the ripple profile as an increase of surface traction from the point of inflexion between trough and crest to the ripple crest. Such a pattern is broadly consistent with the variation in creep transport rate required for the ripple geometry to be preserved as it advances downwind.

Vertical momentum change, on the other hand, becomes greater with increasing θ. Thus, the vigour with which saltations are initiated is greater on the upwind face of the ripple than it is at the crest. The fluctuations in creep and in saltation activity appear to be mutually displaced by approximately half the ripple wavelength.

It should be noted that this fluctuation in the initial conditions for saltation, with the zone of greater vigour upwind on a particular ripple, will lead to an uneven distribution of collisions in the area downwind in which these saltations terminate. This suggests a mechanism for the development of a ripple field from a single surface feature, and also emphasizes the naivety of the process (assuming a uniform distribution of collision) which led to the "stress" figures in rows 4 and 5 of Table 5. Clearly, there are several features to be clarified before a reliable model can be attained of the interaction between saltation and the bed condition.

Because mean creep velocity is very much lower than the mean forward velocity in saltation, while it is known that creep accounts for approximately one quarter of the bed load transport rate, it is clear that a large proportion of the grains in motion at a given time are creeping. Thus, the creep transport rate is a good register of the degree of surface agitation. As has been seen, the greatest rates of creep, and therefore of surface disturbance, occur at the ripple crest. It is there, consequently, that the winnowing out of fine material would be expected to be most active, and the surface layer coarsest.

Conclusions

Grain shape has a pronounced effect, not only on the kinetics of the saltation path, but also on the intervening collisions. Collision is a less effective mechanism for sustaining saltation of shell (platy) sand than of quartz (compact) sand.

The loss of forward momentum per collision evident from the behaviour of grains approaching and leaving the bed is less for shell than for quartz sand. More forward momentum is therefore transferred by the latter to creeping grains.

Much greater change of vertical momentum occurs, on average, at quartz grain collisions than at corresponding shell grain ones. Collision alone can, therefore, support a much smaller load of saltating shell grains than quartz grains.

Adverse bed slope such as occurs on the upwind surface of a ripple, reduces the change of forward momentum which occurs at collision, but increases the change of vertical momentum. In both cases, the sensitivity to bed slope is greater for coarse fractions than for fine, and is different for quartz and shell sands.

It can be inferred from these variations with bed slope that creep activity increases along the ripple back as the crest is approached, but that the vigour with which new saltations are launched is greater near the point of inflexion and decreases as the crest is approached. The greater creep activity at the crest should lead to a coarser grain population there than elsewhere on the ripple.

Acknowledgements

The study is supported by the Natural Environment Research Council. Laboratory and computation facilities were provided by the University of Aberdeen.

References

Bagnold, R.A. 1941. *The physics of blown sand and desert dunes.* London: Methuen.

Hunt, J.C.R. and P. Nalpanis In press. Saltating and suspended particles over flat and sloping surfaces. *Proceedings, International Workshop on Physics of Blown Sand. Memoirs 8.* Department of Theoretical Statistics, Aarhus University.

Jensen, J. and M. Sorensen 1983. On the mathematical modelling of aeolian saltation. In *Mechanics of Sediment Transport*, B. Mutlu Sumer and A. Muller (eds.), 65-72. Rotterdam: Balkema.

Rumpel, D.A. 1985. Successive aeolian saltation: studies of idealized collisions. *Sedimentology* 32, 267-80.

Sneed, E.D. and R.L. Folk 1958. Pebbles in the Lower Colorado River, Texas. A study in particle morphogenesis. *Journal of Geology.* 66, 114-50.

Tsuchiya, Y. and Y. Kawata 1972. Characteristics of saltation of sand grains by wind. *Proceedings 13th International Coastal Engineering Conference* 1617-25.

Willetts, B.B. and M.A. Rice In press a. Wind tunnel tracer experiments using dyed sand. *Proceedings of International Workshop on Physics of Blown Sand. Memoirs 8.* Department of Theoretical Statistics, Aarhus University.

Willetts, B.B. and M.A. Rice In press b. Inter saltation collisions. *Proceedings of International Workshop on Physics of Blown Sand. Memoirs 8.* Department of Theoretical Statistics, Aarhus University.

Willetts, B.B. and M.A. Rice In press c. Collisions in aeolian saltation. *Acta Mechanica.*

White, B.R. and J.C. Schulz 1977. Magnus effect in saltation. *Journal of Fluid Mechanics* 81, 497-512.

2

Small scale wind tunnel modelling of particle transport — Froude number effect

James D. Iversen

Abstract

The aeolian wind tunnel is a useful instrument for studying saltation physics as well as for modelling drift geometry at small scale. For small-scale modelling, it is shown that, although the dimensionless deposition or deflation rate may be a function of several other dimensionless parameters, the Froude number is one of the more important ones. Examples are presented illustrating the dependence of dimensionless deposition or deflation rate or drift geometry on Froude number, density ratio, and the ratio of wind speed to threshold wind speed.

Introduction

The small-scale modelling of sand-transport phenomena is indeed a complex area of study. Wind tunnel modelling of control structures has sufficient advantages over field testing to warrant some discussion of the realism of such experiments. Control of experimental factors such as wind direction and duration is a primary advantage. It is also advantageous to be able to perform many experiments over a short period of time and even to alter particle properties by using different grain materials.

Maintaining similitude for modelling at small scale the effects of blowing particles of dust, sand, or snow is a problem because of the large number of variables. It is important to arrange the variables in dimensionless groups (Iversen 1982). In a true model,

these parameters must all have the same value in the model as in
full scale, but because this is impossible, it is necessary to abandon
the attempt at a true model. The only way it is possible to obtain
realistic quantitative (and perhaps also qualitative) results for a
distorted model is to vary the degree of distortion in order to
facilitate extrapolation to full scale. The degree of distortion is
varied by changing the experimental values of the dimensionless
parameters as much as possible by changing particle density and
diameter, wind speed, and model scale. In addition, interpretation
of results is aided by grouping the dimensionless parameters by
theoretical means in order to reduce the number of variables.

Transport rate similitude

The Mariner 9 spacecraft, which began orbiting the planet Mars in
1971, revealed the presence of many surface craters that possess
dark streaks extending in the leeward direction. Many of these
streaks are probably caused by deflation of small particles resulting
from increased shear stress in the crater wake.

Experimental correlation of gross erosional and depositional
features near model craters in a wind tunnel was obtained (Iversen
et al. 1975) by basing a similitude on rate of mass movement. The
transport rate similitude is based on the particle mass transport
rate,

$$q \sim \frac{\rho}{g} u_*^2 \, (u_* - u_{*t}) \tag{1}$$

Equation 1 is Lettau and Lettau's (1978) version of Bagnold's (1941)
equation.

If it is assumed that Equation 1 also holds for a deposition or
deflation rate q_d, then the equation can be written in terms of the
drift plan area A_d, volume V, or cross-sectional area A_c:

$$\frac{d\left(\dfrac{A_d}{L_r^2}\right)}{d\left(\dfrac{u_*t}{L_r}\right)}, \ \frac{d\left(\dfrac{V}{L_r^2 H}\right)}{d\left(\dfrac{u_*t}{L_r}\right)}, \text{ or } \frac{d\left(\dfrac{A_c}{L_r H}\right)}{d\left(\dfrac{u_*t}{L_r}\right)} \sim \frac{\rho}{\gamma}\left(\frac{u_*^2}{gH}\right)\left(1 - \frac{u_{*t}}{u_*}\right) \tag{2}$$

Effect of Froude number and density ratio

Owen (1964) shows that the equivalent roughness height in
saltation, z_0', is proportional to u_*^2/g; i.e.,

$$z_0' \sim \frac{u_*^2}{2g} \tag{3}$$

If the roughness height - characteristic length ratio z_o'/H is important, then the Froude number u_*^2/gH is also important, according to Equation 3.

Consider a case in which deposition is governed by a topographic feature of height H and lateral width L_r (such as a rock or leeward- or windward-facing escarpment). The volumetric rate at which material is deposited should be proportional to the product of the excess above threshold of wind speed (or friction speed), volumetric concentration, and frontal area. The maximum mass concentration possible in saltation is usually considered to be the fluid density ρ, so that the volumetric concentration is ρ/γ. Thus,

$$dV/dt \sim (\rho/\gamma)HL_r(u_*-u_{*_t}) \qquad (4)$$

and the mass deposition rate is:

$$q_d L_r = \gamma dV/dt \sim \rho HL_r (u_*-u_{*_t}) \qquad (5)$$

From Equation 1, the ratio of deposition rate to total mass flow rate is thus,

$$q_d/q \sim \rho HL_r u_*/(\rho u_*^3 L_r/g) = gH/u_*^2$$

$$(6)$$

or,

$$q_d \sim q/(u_*^2/gH)$$

so that a dimensionless deposition rate would be expected to decrease with an increase in Froude number.

Experiments with different values of particle density have shown the deposition rate to be a function of density ratio as well (Iverson 1982). If the particle concentration within the deposition region is volume limited rather than mass limited, then the deposition rate becomes:

$$q_d \sim q/(\rho u_*^2/\gamma gH) \qquad (7)$$

A series of wind tunnel experiments was performed in a manner similar to the model crater experiments to determine the snowdrift deposition rate associated with an interstate highway grade separation structure (Iversen 1980b). Values of friction speed were estimated from the wind speed measurements using a logarithmic velocity profile. The data for mass deposition rate from ten experiments were curve-fitted by linear regression to compare Equations 6 and 7. The results are as follows:

$$q_dg/\rho u_*^3 = 1.4\ (u_*^2/gH)^{-0.49},\ R = 0.74 \tag{8}$$

$$q_dg/\rho u_*^3 = 0.064\ (\rho u_*^2/\gamma gH)^{-0.41},\ R = 0.90 \tag{9}$$

It is clear that the correlation (R = correlation coefficient) with densimetric Froude number (Eqn 9) is better than with the ordinary Froude number (Eqn 8). However, the dimensionless deposition rate is not as strong a function of Froude number or densimetric Froude number as predicted by Equations 6 or 7. By multiple regression, treating the Froude number and density ratio separately, the best fit was found to be:

$$q_dg/\rho u_*^2\ (u_* - u_{*t}) = 0.0696\ (\rho/\gamma)^{-2/5}\ (u_*^2/gH)^{-7/9},\ R = 0.97 \tag{10}$$

The form of the best fit to any set of data is undoubtedly a strong function of the geometry of the boundary layer obstruction. Since only a very few experiments of this type have yet been performed, much work remains to establish the mass transport similitude on a firmer foundation. Equation 10 is used in plotting the data as shown in Figure 1. Full-scale values of the density ratio-Froude number parameters are approached at the left side of the figure.

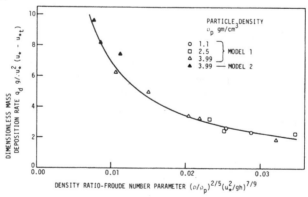

Figure 1 Rate of deposition of simulated snow on a model roadway, as a function of density ratio–Froude number parameter (Iversen 1982).

The similitude function

The deposition or deflation rate function, Equation 2, provides a basis for analysis of drift control simulation. As shown in Figure 1, the drift transport rate can be a function of more than one parameter. More generally,

$$\frac{d(A_d/L^2)}{d(u_*t/L_r)} = \frac{\rho u_*^2}{\gamma gH} \ f \left\{ \frac{u_*}{u_{*t}} \ , \ \frac{u_*^2}{gH} \ , \ \frac{z_o}{H} \ , \ \frac{z_o'}{H} \ , \ \frac{U_F}{u_{*t}} \ , \ \frac{u_*L}{\nu} \ \ \frac{\rho}{\gamma} \ , \ \text{geometry} \right\}$$

(11)

Equation 11 allows for the possibility of vertical geometric distortion and uses Jensen's (1958) criterion for normal roughness modelling. It does not allow for the satisfaction of many of the original dimensionless terms. Thus, the saltation model is still highly distorted. By varying particle size and density, model speed, scale, and possibly vertical distortion, a range in degree of distortion can result so that full-scale predictions can be made. The effects of distortion become particularly serious, however, if small-scale features such as surface ripples become large enough to obscure or interfere with gross drift geometry associated with important topographic features of the model.

It is assumed in Equation 11 that the velocity profile is that for neutrally stratified air. If it is not neutrally stratified, then another parameter must be included to account for a non-neutral profile. The wind speed profile for neutral stratification is controlled by topographic geometry and surface roughness z_o (and z_o' in saltation). Since it is easier to measure wind speed than surface friction speed, it may be convenient to replace the u_* and u_{*t} in most of the factors in Equation 11 with the reference wind speeds at reference height H, i.e. U and U_t. For obstructions having height only a fraction of the saltation layer depth, the density ratio ρ / ρ_p seems to be one of the important separate functional parameters.

Accumulation rate calculation and experimental comparison

A controversy (Anno 1984, Iversen 1984, 1985) concerning prediction of full-scale drift phenomena from small-scale model tests arises because the model is distorted. Tabler (1980b) and others (e.g. Kind 1976) have felt that the Froude number is the single most important parameter governing drift conformation. Anno, however, states that the ratio of wind speed to the wind speed at threshold is more important.

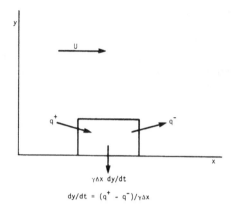

Figure 2 Mass balance of a segment of a saltation layer in a region of streamwise-dependent surface friction speed. The rate of accumulation of material is proportional to the net rate of transport q^+-q^- into the segment.

In order to assess the relative importance of the Froude number and the speed ratio, a series of calculations was carried out in order to approximate the initial deposition and erosion rates adjacent to a drift-control barrier. Figure 2 illustrates the mass balance of a short saltation layer segment of streamwise length X. The amount of mass per unit time per unit lateral dimension entering the segment is q^+, the amount leaving the segment is q^-, and the net rate of accumulation or deflation of material on the surface is proportional to the difference.

$$dy/dt = (q^+ - q^-)/(\gamma \Delta X) \tag{12}$$

It is assumed that the mass transport rate q^- leaving the segment is simply determined by the local shear stress as given by Equation 1. The transport rate entering the segment, however, can be influenced by conditions far upstream. This effect is approximated by calculating the path length, so that material entering the segment is assumed to leave an area which is one path length upwind (see Phillips & Willetts 1979, Jensen 1959). Owen (1980) hypothesized that the path length is proportional to the Froude number, i.e.

$$L = 0.37D(U_F/u_*)u_*^2/g \tag{13}$$

where L is the path length and D is the ratio of the wind speed within the saltation layer to the friction speed.

An example of the calculation procedure is shown in Figure 3. The region shown from point 0 to point 19, which is 19 segment lengths downwind, is a region of decreasing shear stress such that at point 0, the friction speed u_* is equal to the uniform upwind value $u_{*\infty}$, and has decreased to 74% of that value at point 13. The corresponding calculations of path length for a Froude number $u_{*\infty}^2/gH = 1$, range from 10.3 times the segment length (equal here to the obstacle height H) at point 1 to 5.6H at point 13. Calculations of the net deposit rate $\Delta q/q_\infty$ range from zero upwind of point 1 to 0.10 in Bin B to 1.05 in Bin L, which, because of the reduction in path length from point 1 to point 3, receives 83% of the particles leaving Bin B and 83% of the particles leaving Bin C. The value of q^- for each bin is taken to be the average value.

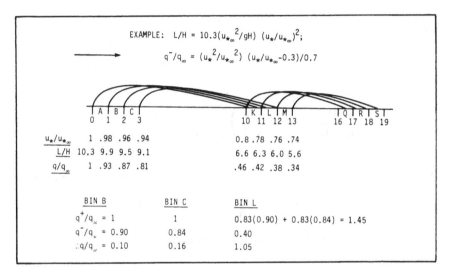

Figure 3 Calculation procedure for determination of accumulation rate in a region of streamwise-dependent surface friction speed. The material entering Bin L arrives from Bins B and C. The entering transport rate q^+ in Bin L is thus a function of the value of u_* in Bins B and C. The leaving transport rate q^- from Bin L is assumed to be a function of the friction speed at Bin L. The net rate of accumulation is proportional to $\Delta q = q^+ - q^-$.

This is, of course, a rather approximate calculation because the method does not properly model the physics of the saltation process in a region of streamwise-varying shear stress. It should give some qualitative insight, however, into the effect of both parameters in question, i.e. Froude number (contained in the particle path length calculation) and friction speed ratio (contained in Eqn 1 for the calculation of q).

A mathematically expedient distribution of surface friction speed is shown in Figure 4. This distribution has minimum values of surface friction speed located at x/H = -5 (80% of free stream) and at x/H = +10 (60% of free stream) and a maximum (102.5% of free

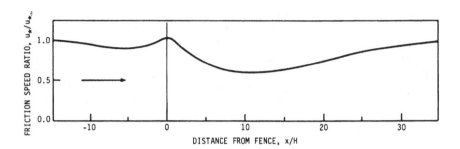

Figure 4 Assumed streamwise surface friction speed distribution near a porous barrier with a bottom gap (located at x/H = 0).

stream) at $x/H = 0$, simulating a surface gap underneath a porous fence.

The rate of deposition of erosion is given by Equation 12. A dimensionless rate of accumulation can thus be written (using Eqn 1):

$$\frac{d(y/H)}{d(tu_{*_\infty}/H)} \sim \frac{\Delta q}{q_\infty} \left(\frac{\rho}{\gamma}\right) \frac{H}{\Delta x} \left(\frac{u_{*_\infty}^2}{gH}\right)\left(1 - \frac{u_{*t}}{u_{*_\infty}}\right) \qquad (14)$$

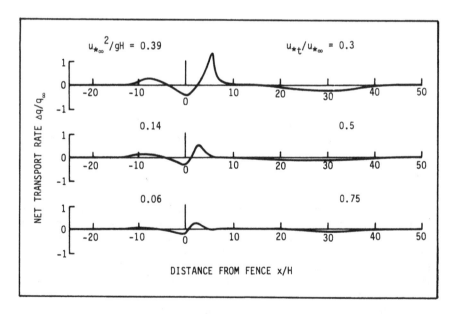

Figure 5 Net accumulation rate as a function of distance from the barrier for three sample calculations using the procedure of Figure 3. Note that the location of maximum accumulation rate moves downwind as the Froude number $u_{*_\infty}^2/gH$ increases.

As soon as a drift begins to form, the surface topography, and therefore the surface friction speed distribution begin to change. If the calculation of drift geometry is to proceed, these changes must be taken into account. Here, we assume that the drift formation is just beginning. Typical results are shown in Figure 5 for three values of the ratio $u_{*_t}/u_{*\infty}$. For constant barrier height and threshold friction speed, this corresponds to three different values of Froude number as shown. Similar calculations were made for other combinations of friction speed ratio and Froude number. Figure 5 shows that the location of maximum accumulation varies with the values of the parameters. Results from the entire set of calculations are illustrated in Figures 6, 7, and 8. Figure 6 shows that the distance downstream of the location of maximum accumulation increases with the Froude number (corroborated in Fig. 5 of Iversen 1982). The crossplots of Figures 7 and 8 show, on the other hand, that neither the location of maximum accumulation rate nor the value of $\Delta q/q_\infty$ varies with friction speed ratio except for values of friction speed very near threshold.

Some experimental evidence for the effect of Froude number on drift geometry that appears to justify these results is shown in Figures 9-12. Photographs of the drift geometry associated with a 50% porous barrier fence with a 10% bottom gap are presented in Figure 9. The characteristics of these drifts and others are tabulated in Table 1 and plotted as functions of the Froude number in Figures 10-12. The experimental location of the leeward drift leading edge is illustrated in Figure 10. The experimental trend toward increase of distance downwind with Froude number is the same as the calculated result shown in Figure 6. It is interesting to note that the downwind location of the maximum leeward drift depth exhibits a slight opposite trend from the leading edge location as shown in Figure 11. The value of maximum depth appears to decrease slightly with Froude number, as Figure 12 illustrates.

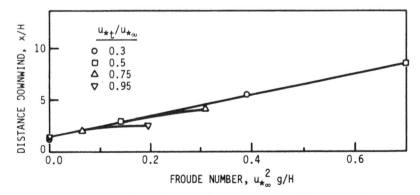

Figure 6 Variation of the location of maximum accumulation rate downwind of the barrier as a function of Froude number. A Froude number of zero is obtained by setting particle trajectory length to zero. Calculation procedure of Figure 3.

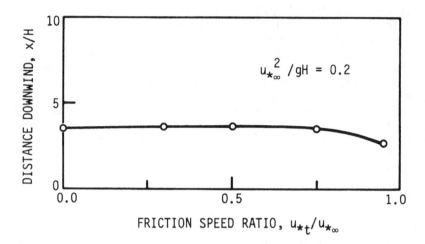

Figure 7 Variation of the location of maximum accumulation rate downwind of the barrier as a function of friction speed ratio for constant Froude number. Calculation procedure of Figure 3.

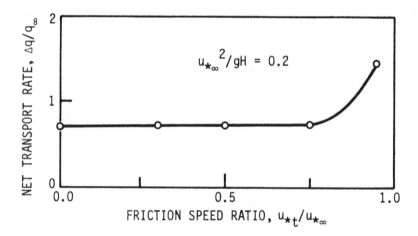

Figure 8 Variation of the value of maximum accumulation rate downwind of the barrier as a function of friction speed ratio for constant Froude number. Calculation procedure of Figure 3.

Conclusions

The wind tunnel can be a very useful tool for the analysis and design of windbreak and sand drift control devices, not only for measuring wind speeds and turbulence intensities in the normal wind tunnel role, but also for modelling at small scale the dust and sand transport associated with such devices. For the latter purpose, the results of the foregoing rather preliminary analysis

Figure 9 Photographs of full scale and model Fence barriers (50% porosity, 0.1H bottom gap).**A** Full scale snow fence (3.8m) adjacent to Interstate 80, Wyoming, USA.**B** Wind tunnel fence model (Iversen 1980a, $U_h^2/gH = 0.14$). **C** Wind tunnel fence model (Iversen 1981, $U_h^2/gH = 0.16$). **D** Frozen lake model, Wyoming (Tabler 1980a, $U_h^2/gH = 0.044$). Photograph courtesy of Ronald D. Tabler. **E** Frozen lake model, Wyoming (Tabler 1985, $U_h^2/gH = 0.28$). Photograph courtesy of Ronald D. Tabler. **F** Water channel model (Wuebben 1978, $U_h^2/gH=0.015$), Photography courstesy of James L. Wuebben.

seem to indicate that the Froude number is a more important parameter for determining drift conformation than is the speed ratio $u_{*t}/u_{*\infty}$. The experimental trend of the downwind drift leading edge location as exhibited in Figure 10 corroborates the Froude number effect. It also appears necessary, however, to perform experiments at speeds perhaps at least 20% above threshold in order to prevent the speed ratio from having a significant effect. In any event, at small scale, a distorted model is extant, and it is necessary to perform an extensive series of experiments if quantitative extrapolation to full scale is desired.

Table1 Drift geometry for a porous fence barrier.

Source	Fence Height (cm)	Particle Material	Froude Number $u_{*\infty}^2/gH$	Location Drift Windward Edge x/H	Location Maximum Drift Depth x/H	Maximum Drift Depth z/H
Tabler (1985)	12	snow	0.280	3.6	4.1	0.80
Kind & Murray (1982)	10[a]	sand	0.190	_[b]	1.8	0.96
Iversen[c] (1980)	2.5	glass	0.160	2.1	4.8	0.77
Iversen (1981)	2.5	glass	0.140	1.3	4.2	1.10
Tabler (1980a)	12	snow	0.044	0.0	5.0	1.15
Kind & Murray (1982)	10	polystyrene	0.030	_[b]	2.4	0.89
Tabler (1980b)	378	snow	0.020	0.0	6.6	1.21
Wuebben (1978)	2.4	sand (in water)	0.015	_[b]	3.3	1.19

[a]Model fence of approximately 60% porosity.
[b]Continuous drift from windward to leeward of fence.
[c]Drift geometry only 80% capacity.

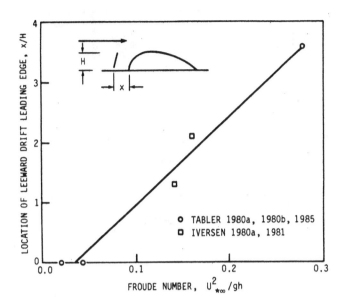

Figure 10 Experimental values of the windwise location of leeward drift leading edge for 50% porosity fence barriers as a function of Froude number. Note the similarity with the approximate calculation of the location of initial drift deposit shown in Figure 6.

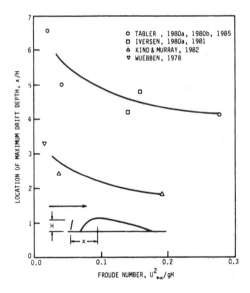

Figure 11 Experimental values of the windwise location of maximum depth of the leeward drift for porous fence barriers as a function of Froude number. The difference between the upper and lower curves may be because the three data points near the lower curves are from experiments for which the drift deposit from the windward drift to the leeward drift was continuous through the fence barrier.

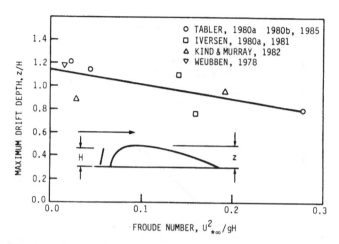

Figure 12 Experimental values of the leeward maximum drift depth for porous fence barriers as a function of Froude number.

Acknowledgements

The research reported in this paper has been supported by the U.S. National Aeronautics and Space Administration, the Iowa Department of Transportation, and by the Engineering Research Institute, Iowa State University.

References

Anno, Y. 1984. Applications of Anno's modelling conditions to outdoor modelling of snowdrifts. *Cold Regions Science & Technology* 9(2), 179-81.

Bagnold, R.A. 1941. *The physics of blown sand and desert dunes.* London: Methuen.

Iversen, J.D. 1980a. Drifting snow similitude--transport rate and roughness modelling. *Journal of Glaciology* 26, 393-403.

Iversen, J.D. 1980b. Wind tunnel modelling of snow fences and natural snow fence controls. *Proceedings, Eastern Snow Conference*, 106-24.

Iversen, J.D. 1981. Comparison of wind-tunnel model and full-scale fence drifts. *Journal Wind Engineering and Industrial Aerodynamics* 8, 231-49.

Iversen, J.D. 1982. Small-scale modelling of snow drift phenomena. In *Wind tunnel modelling for civil engineering applications.* 522-45. Cambridge: Cambridge University Press.

Iversen, J.D. 1984. Comparison of snowdrift modelling criteria. *Cold Regions Science and Technology* 9, 259-65.

Iversen, J.D. 1985. Wind tunnel modelling of sand transport physics and control. *Proceedings, United National International Workshop on Sand Transport & Desertification in Arid Lands.* Khartoum, Sudan.

Iversen, J.D., R. Greeley, B.R. White and J.B. Pollack 1975. Eolian erosion of the martian surface, Part I: Erosion rate similitude. *Icarus* 26, 321-31.

Jensen, M. 1958. The model-law for phenomena in natural wind. *Ingeniøren* 2, 121-8.

Jensen, M. 1959. *Aerodynamik i den Naturlige Vind.* Teknisk Forlag, København, Denmark.

Kind, R.J. 1976. A critical examination of the requirements for model simulation of wind-induced erosion/deposition phenomena such as snow drifting. *Atmospheric Environment* 10, 219-27.

Kind, R.J. and S.B. Murray 1982. Saltation flow measurements relating to modelling of snowdrifting. *Journal Wind Engineering and Industrial Aerodynamics* 10, 89-102.

Lettau, K. and H.H. Lettau 1978. Experimental and micro-meteorological field studies of dune migration. In *Exploring the world's driest climate*, H.H. Lettau and K. Lettau (eds.), 110-147. University of Wisconsin-Madison, Institute for Environmental Studies, IES Report 101.

Owen, P.R. 1964. Saltation of uniform grains in air. *Journal of Fluid Mechanics* 20, 225-42.

Owen, P.R. 1980. *The physics of sand movement. Workshop on physics of desertification.* Trieste, Italy: International Centre for Theoretical Physics.

Phillips, C.J. and B.B. Willetts 1979. Predicting sand deposition at porous fences. *Journal of the Waterway, Port, Coastal and Ocean Division, Proceedings ASCE* 105, WWI, 15-31.

Tabler, R.D. 1980a. Geometry and density of drifts formed by snow fences. *Journal of Glaciology* 26, 405-19.

Tabler, R.D. 1980b. Self similarity of wind profiles in blowing snow allows outdoor modelling. *Journal of Glaciology* 26, 421-34.

Tabler, R.D. 1985. Personal communication.

3

Mass momentum and kinetic energy fluxes of saltating particles

Dale A. Gillette and Paul H. Stockton

Abstract

New instrumentation is used for measuring mass flux and momentum flux near the leading edge of an erodible deposit of particles in a wind tunnel. Measurements changed greatly as a function of distance from the leading edge of the particle deposit. The changes were shown to derive from changes in friction velocity resulting from a particle-wind speed interaction and from a wind tunnel constraint. The sand flux and wind profile vertical momentum flux affect each other in a kind of feedback system. Beyond a distance of 1 m from the leading edge of the erodible particle deposit, the ratio of the vertical kinetic energy flux to the horizontal mass flux was almost constant.

Introduction

We designed an experiment to evaluate some relevant parameters of movement for saltating particles as a function of length of particle deposit and wind speed in a wind tunnel. These parameters - mass flux, momentum flux, and kinetic energy flux - specify particle flow, friction of the particle-air mixture, and sandblasting kinetic energy flux. A new instrument was built to measure momentum flux of particles for this experiment. In evaluating the above fluxes, airborne particle mass concentration profiles and particle speeds were also estimated.

Since in our experiment the particle motion was designed to be dominantly saltation, the particle kinetic energy is very important to the potential mass abraded by sandblasting. Greeley *et al.* (1983) showed for several materials that the mass of material abraded per impact of a sand grain is directly proportional to the kinetic energy of the impacting particle. Hagen (1984) also showed that the mass of soil aggregate material abraded by impacting particles increased as a power of the velocity of the abrading particle. The powers were from 1.5 to 2.3 for the stable aggregates and near 2.0 for fragile aggregates. For abrasion of objects above the ground such as rocks (Greeley *et al.*, 1983), the horizontal flux of mass, momentum, and kinetic energy of particles is of primary interest. For abrasion of the soil, however, the vertical flux of momentum and kinetic energy of particles is of primary interest. This investigation is concerned with both horizontal and vertical fluxes.

Owen and Gillette (in press) described the wind stress of the wind tunnel used in our experiment as a function of length of particle deposit, particle size, wind speed and wind tunnel roof height. Here, we describe the complementary particle movement measurements obtained and their relationship with wind and deposit length.

Experimental Procedure

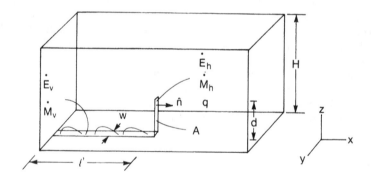

Figure 1 The arrangement of the control volume in the wind tunnel and the illustration of parameters (see text for definitions).

Control volume and definitions

The experiments used a small suction-type wind tunnel having a working cross section 15.5 cm wide x 19.5 cm high, which was very similar to that described by Gillette (1978). The control volume within the working section of the wind tunnel is pictured in Figure 1 along with many of the parameters used in this paper.

Definitions and units used in Figure 1 and used frequently in the paper follow:

w = width [cm] of control volume

d = height of internal boundary layer containing particles [cm]

ℓ = length to leading edge of deposit (deposit length) [cm]

u = wind speed = $\vec{u} \cdot \hat{n}$ [cm s^{-1}]

V = particle speed = $\vec{v} \cdot \hat{n}$ [cm^{-1}]

U = mean wind speed = $\int_0^H u(z)dz/H$

H = height of wind tunnel [cm]

c = concentration of particles in the air [g cm^{-3}]

ρ = density of air [g cm^{-3}]

q = horizontal mass flux through surface A [g cm^{-1} s^{-1}]

$\dot{M}_h$ = horizontal particle momentum flux through surface A [dynes cm^{-1}]

$\dot{M}_v$ = vertical momentum flux from particles striking the floor surface of width w, length ℓ [dynes cm^{-1}]

$\dot{E}_h$ = horizontal kinetic energy particle flux through surface A [ergs cm^{-1} s^{-1}]

$\dot{E}_v$ = vertical flux of particle kinetic energy striking floor surface of length ℓ width w [ergs cm^{-1} s^{-1}]

$\hat{n}$ = unit vector parallel to wind direction

T = time duration of experiment

Momentum flux measurement

A new instrument was built to measure the momentum flux of particles directly. This option was chosen rather than calculating momentum flux from the measurement of concentrations and speeds of individual particles. Although individual particle speed was measured by White and Schulz (1977) and by Schmidt (1977), we desired to measure momentum flux at many heights consisting of the movements of thousands of particles including those very close to the ground where the thick cloud of particles obscures single particle identification (White & Schulz 1977). We also wished to bypass the need for particle-by-particle calculation. An instrument was desired that would respond only to particle momentum, not to changes of wind speed, would respond linearly with particle momentum increase, would be stable in its response, and would not respond to impacts of particles except on its own target area. In addition, a profile of momentum flux was desired having at least 1-cm resolution for a 10-cm-deep layer above the ground surface.

To meet these requirements, an impactor was constructed whose impaction surfaces are covered piezoelectric crystals; the response of these crystals to impact by sand-sized particles is amplified and recorded.

Sensors   The sensors consist of an array of 10 ceramic crystals. Each sensing surface is a brass plate, approximately 2 mm wide and 9.5 mm high, laminated to each crystal. The sensors are vertically oriented to provide an integrated total momentum profile versus height for each centimeter interval of height from 0 to 10 cm.

The sensing head is shaped to produce a minimum disturbance to the air flow (see Fig. 2). The sensing surface is flat to simplify as much as possible the errors associated with lost energy due to nonperpendicular impacts upon the sensing surfaces. It is as narrow as possible to reduce flow disturbance. The angle between the sensing surfaces and the side of the head is approximately 96 degrees. Within the head are high-impedance preamplifiers to provide an amplified low-impedance signal to the main signal-processing unit.

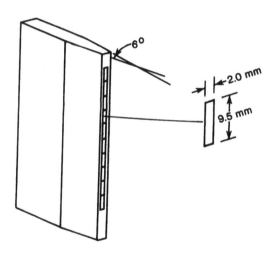

Figure 2 The physical arrangement of the 10 momentum-sensing impactors vertically stacked and encased with high impedance preamplifiers.

Circuit Description   The voltage developed across each crystal is proportional to the momentum impulses imposed upon the crystal as particles impact the sensing surface. The high-impedance amplifiers provide some gain to this signal, and are selected primarily to solve several problems. The most essential function is to follow the voltage of the crystal with a high enough impedance that no appreciable current is drawn from the crystal. The total momentum is proportional to the area under the voltage curve generated by the crystal as particles impact upon the sensing surface.

The signal from the sensing head is transmitted to the analog processing unit by a shielded cable (see Fig. 3). The first high-pass filter serves to remove low-frequency fluctuations caused by temperature drift of the crystals, by preamplifiers, and predominantly by the unwanted signal from the momentum of the wind. A band-pass filter then helps remove HF noise from the amplifiers. To keep the analog integrators very stable, a minimum threshold of acceptable signal is determined by passing the signal through an extremely high reverse-impedance CMOS diode. This threshold level is internally adjustable.

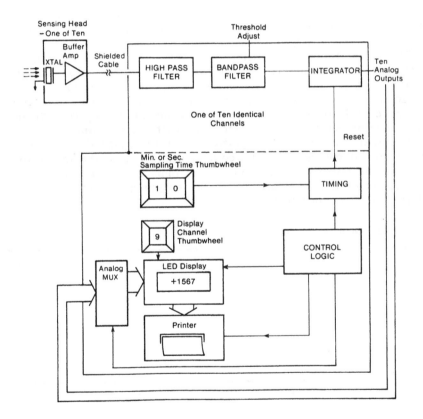

Figure 3 A block diagram of the particle momentum profile instrument.

Current through the diode is proportional to the voltage of the signal. This current is summed in a classic active inverting integrator. The integrator's capacitor charge builds as $Q = (IT')$, where I is the average current for a period of time T'. The integrator amplifier's output is a negative voltage proportional to the sum of the charging currents $V_c = Q/C'$, where C' is the integrator's capacitance and V_c is the voltage developed across the capacitor. The instantaneous voltage signal V_o is fed into an integrator, and the output voltage V_o is given by:

$$V_o = \int_0^T V_{in}\, dt.$$

The output voltages represent a 10-point profile of impacted momentum versus height.

The integration time is selectable from 1 second to 99 minutes. After data are accumulated over the selected time period, the ten analog voltages are digitized and printed, the integrators are reset to zero, and sampling resumes for another time period. A digital voltmeter will display the value for any channel at any time during the sampling period without interfering with the sampling.

The analog method of integration is an advantage if the impacts are frequent enough to occur coincidently, because the integration will be the response to the sum of the impacting particles. Digital processing of the waveform has data analysis advantages, however, if the time for processing is available.

Table 1 Description of glass spheres used in calibration.

	Sphere 2	Sphere 3
Density (g cm^{-3})	1.5 ± 0.05	2.5 ± 0.05
Mass weighted mean diameter (μm)	185.1	271.1
Standard deviation (μm)	30.4	23.4
Hardness (Mohs scale)	6.0	6.0
Terminal sedimentation velocity (cm s^{-1})*	142.0	210.0
Threshold friction velocity	30.0	33.0

* Computed from Equation 1 for altitude of Boulder, Colorado.

Calibration   Calibration of the momentum sensor consisted of injecting glass spheres, for which size, particle speed; and mass flux had been determined, into a vertical tube in which a steady-state wind speed was measured. The tube was aimed directly at the sensor and was separated from it by a 5 mm gap. Characteristics of two different sets of glass spheres that were normally distributed with size are given in Table 1. A third sphere set, sphere 1, was used, but its contents were found to go into direct suspension; consequently it was not used for this paper. The spheres were accelerated from the resting point by the wind and by gravity to the location of the sensor. Wind speed was measured in the calibration tube by a calibrated Pitot tube anemometer. The particle speed V was obtained by solving the equation of motion (see Fuchs 1964):

$$\frac{d(V-U_o)}{dt} = -\frac{3c_D\rho(V-U)|V-U_o|}{8r\,\rho_p} + g \qquad (1)$$

where

U_o is (constant) wind velocity

r is particle radius

ρ_p is particle density

ρ is air density

t is time

g is gravitational acceleration

c_D is the drag coefficient.

The drag coefficient was determined by interpolation from tables of c_D vs. Reynolds number for spheres. An initial Reynolds number was computed for V = 0, t = 0. Equation 1 was solved for constant c_D, and the solution was used to determine the position where the calculated Reynolds number gave $\Delta c_D/c_D$ = 0.01. At that position, a new drag coefficient was used and Equation 1 was solved for V and a new position for which $\Delta c_D/c_D$ = 0.01. The particle speed was thus found iteratively until the position of the sensor was reached. Usually 50 to 100 iterations were required to solve for particle speed at the position of the sensor. The calculated particle velocity was compared with measurements of particles in the calibration apparatus by optical scintillation (Wang et al. 1977) by the NOAA Wave Propagation Laboratory (WPL). The agreement of calculated and measured particle speed is shown in Figure 4.

When the electronics were tuned to give highly sensitive responses to actual impact, but not to vibrations caused by air flow around the sensor head, a small fraction (0% to 5%) of the impacts on one sensor pad would be additively felt by adjacent sensor pads. To correct for this small amount of "cross-talk" between the channels, the linear equation was written:

$$Y_i = \Sigma_j a_{ij} X_j \underset{j}{\Sigma}$$

where Y_i is the response for channel (sensor) i, a_{ij} is a multiplicative constant expressing the ratio of response of channel i to that of channel j (X_j) when only channel j is struck. To solve for the actual response X_i from the apparent response Y_i, the equation $Y = AX$ was inverted. Values of a_{ij} were calculated by measuring the response of channel i to impacts on channel j when all channels except channel j were covered by several layers of rubber sheets. Typical values of the corrected response X_i from the uncorrected response Y_i differed by a maximum of about 20%. Values for a_{ij} were found for the entire range of particle speeds used in the calibration.

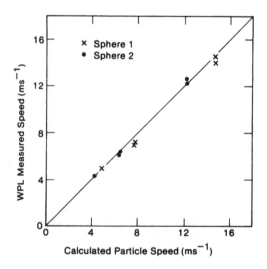

Figure 4 Calculated speed of calibration particle vs. particle speed measured by the NOAA Wave Propagation Laboratory's optical scintillation device.

Gain and level were adjusted for each experiment to better minimize crosstalk and maximize the response for a given sphere. As is shown in Figure 5 this led to a thresholding phenomenon in which a fraction of the signal was lost. To minimize this effect, experiments were designed so that all responses were well above the threshold.

Successive calibrations indicated stability and linearity of instrument response. Calibration mean values of response vs. total momentum flux were not statistically different for two calibrations separated by several months.   For the range of our tests, the sensors did not become saturated.

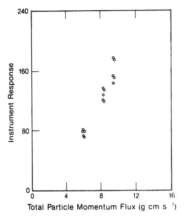

Figure 5 Instrument response vs. total particle momentum flux for the calibration for monodisperse glass spheres.

Mass flux measurement
Mass flux was determined by a Bagnold Catcher. The first such instrument was discussed by Bagnold (1941). Our version of the instrument is identical in shape to the momentum sensor, but collects particles rather than sensing momentum.   Where the momentum sensor has impaction plates, the Bagnold Catcher has holes connected to individual reservoirs in the body of the catcher. Mass fluxes were determined by placing the Bagnold Catcher in the same position relative to the calibration tube and rerunning the calibration for identical conditions.   In a second stage of calibration, the Bagnold Catcher efficiency was determined by comparing the collected material with known mass flux.   The calibration momentum fluxes were then:

$$\frac{V\,m_B}{eff_B}$$

where V is calculated from Equation 1, m_B is the mass flux for the Bagnold Catcher for the calibration run, and eff_B is the efficiency of the Bagnold Catcher.   For our particular Bagnold Catcher we confirmed Horikawa and Shen's (1960) result that the efficiency of this kind of sand trap is almost independent of wind speed.   For the particles used in this experiment, the Bagnold Catcher had an efficiency of 0.70 ± 0.05.

Wind speed measurements
Wind speed measurements for these experiments were reported by

Owen and Gillette (in press). Profiles of wind speed were analyzed in that paper for friction velocities u_* that are given in Table 2 along with mean wind speed U. The accuracy with which u_* could be determined was about $\pm 30\%$.

Table 2 Friction velocities u_* and mean wind velocities U for experiments.

Sphere	Speed	Deposit length (cm)	u_* (cm s^{-1})	U (cm s^{-1})
2	Low	150	88.0	795
		100	78.4	821
		50	65.0	909
		30	46.0	950
		20	41.0	937
		10	35.0	975
	High	150	129.0	1058
		100	119.0	1063
		50	112.0	1107
		30	112.0	1107
		20	88.0	1150
		10	46.0	1170
3	Low	150	95.0	836
		100	80.0	806
		50	74	908
		30	42.4	928
		20	40.6	986
		10	34.0	1020
	High	150	119.0	1013
		100	114.0	1051
		50	96.8	1132
		30	60.0	1138
		20	51.2	1179
		10	54.4	1198

Precision estimates

Precision of the measurements is expressed below as coefficients of variation,   (the standard deviation about the mean divided by the mean of the measurement):

Measurement	γ
Wind speed	0.03
Mass flux of particles	0.08
Momentum flux of particles	0.15
Speed of particles	0.17
Pressure differential	0.12

Method of the experiment

The impactor was mounted on the centerline in the wind tunnel, and monodisperse glass spheres were uniformly deposited on the floor at varying distances upwind of the sensing head of the instrument. A wind was generated in the tunnel exceeding the threshold velocity of the glass spheres. Wind speed reached equilibrium within about 3 seconds. In about 30 seconds, wind erosion removed the deposit of glass spheres. The impactor operated during the entire time of erosion. On following tests, mass fluxes, air velocity profiles, and differential pressures were measured for a similar configuration of the wind tunnel, but with an identically shaped calibrated Bagnold Catcher, Pitot tube anemometers, and static pressure probes, respectively. Glass spheres were examined before and after several of the tests. The spheres were not broken or visibly altered by the tests.

Ratios of sedimentation velocities to friction velocities (see Tables 1 & 2) are between 1.1-4.1 for sphere 2, and 1.8-6.2 for sphere 3. These ratios are within the range for saltation to be the dominant mode of particle motion (Gillette 1981).

Four sets of experiments were run: two wind speeds (high and low) for two sphere sizes (2 and 3). Each set of experiments was run for six different lengths from the leading edge of the erodible particle deposit (abbreviated here-after as deposit length) of 150, 100, 50, 30, 20, and 10 cm.

Profiles of mean particle speeds and mean particle concentrations

A mean horizontal mass flux for the area (O-w, z_1-z_2) may be defined as:

$$\dot{m}(O\text{-}w, z_1\text{-}z_2) = \frac{\int_0^T \int_0^w \int_{z_1}^{z_2} c\, V\, dz\, dy\, dt}{\int_0^T \int_0^w \int_{z_1}^{z_2} dz\, dy\, dt} \tag{2}$$

where: c is particle concentration [g cm^{-3}]

V is particle horizontal speed [cm s^{-1}],

while a mean horizontal particle momentum flux may be similarly defined:

$$\dot{M}(0\text{-}w, z_1\text{-}z_2) = \frac{\int_0^T \int_0^w \int_{z_1}^{z_2} c\, V^2\, dz\, dy\, dt}{\int_0^T \int_0^w \int_{z_1}^{z_2} dz\, dy\, dt} \tag{3}$$

Consequently, the measurement of mean fluxes $\dot{m}$ and $\dot{M}$, enable us to estimate mean particle horizontal speed and particle mass concentration c at distance ℓ for the area $(0\text{-}w) \times (z_1\text{-}z_2)$. That is,

$$\overline{v}(0\text{-}w, z_1\text{-}z_2) = \frac{\dot{M}}{\dot{m}} \tag{4}$$

and

$$c(0\text{-}w, z_1\text{-}z_2) = \frac{\dot{m}^2}{\dot{M}} \tag{5}$$

It must be emphasized that $\overline{v}$ and c are averaged quantities, and do not refer to single particle motion. This distinction is especially important near the floor where particles being lifted (at the beginning of a saltation trajectory) have very small forward velocities, whereas particles moving toward the surface (at the end of a saltation trajectory) have quite considerable forward velocities.

A plot of mean particle horizontal speed versus natural log of height is given in Figure 6 along with wind speed and ln c vs. ln z for sphere 2, high speed, at $\ell = 150$ cm. It was observed that mean particle horizontal speed was slightly less than wind speed for $\ell > 100$ cm and heights greater than 5 cm (ln z > 1.6). The mean particle horizontal speeds for heights smaller than 5 cm for $\ell > 100$ cm, and for all heights for $\ell < 100$ cm were typically smaller than the wind speed at the same height. This would imply that particle acceleration on the average is occurring in this region.

The plots of ln c vs. ln z shown in Figure 6 suggest that a linear relationship exists between the two, at least for the upper measurements. For some of the plots, either an insufficient number of data points existed to draw straight lines for ln c vs. ln z, or curvature of the data points negated a linear relationship. The negative slopes, Γ, of the fitted straight lines ($\Gamma = -\dfrac{d\ln c}{d\ln z}$) are given in Table 3. The equation:

$$ku_* z\, \frac{dc}{dz} = -\Omega c \tag{6}$$

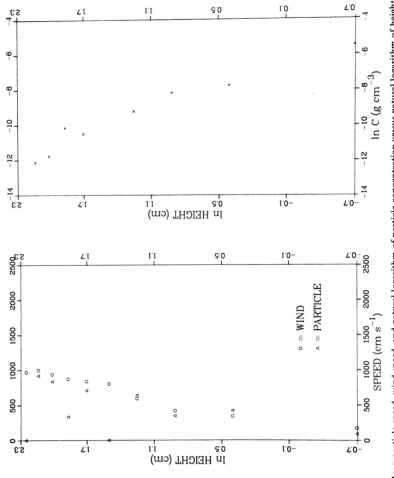

Figure 6 Mean particle speed, wind speed, and natural logarithm of particle concentration versus natural logarithm of height for sphere 2, low-speed experiments at 150 cm from the leading edge of the erodible particle deposit.

where Ω is terminal sedimentation velocity

k is von Karmann's constant

u_* is friction velocity and

z is height

has a solution:

$$\frac{d \ln c}{d \ln z} = - \frac{\Omega}{k u_*} = - \Gamma$$

Thus,

$$u_* = \frac{2.5 \, \Omega}{\Gamma}$$

Using the values for Γ and for terminal sedimentation velocity Ω given in Table 1, we calculated the u_* values and listed them in Table 3 as u_* (fitted). The values of u_* derived from wind profiles from Table 2 u_* (observed) contrasted with the values of u_* (fitted). The fair agreement between u_* (fitted) and u_* (observed) does not confirm the validity of the simplistic assumptions of Equation 6, but rather provides a possibly convenient form to express the concentration change with height.

Table 3 Values of negative Γ and calculated friction velocity u_* for wind tunnel experiments of eroding shperes. (Data unavailable for 10 and 20 cm deposit lengths)

deposit length (cm)	Γ	u_* (fitted) (cm s^{-1})	u_* (observed) (cm s^{-1})	Γ	u_* (fitted) (cm s^{-1})	u_* (observed) (cm s^{-1})
SPHERE 2						
30				4.90	72	102.0
50				4.90	72	112.0
100	3.47	102	78.4	4.90	72	119.6
150	3.51	101	88.0	3.23	110	129.0
SPHERE 3						
30	4.39	120	42.2			
50				4.90	107	96.8
100	4.07	129	80	4.76	110	114.0
150	4.44	118	95	3.56	148	119.0

Taken together, the increase of mean speed with height and decrease of mean particle concentration with height imply differing heights for mass flux, momentum flux, and kinetic energy flux maxima. The height of maximum horizontal kinetic energy flux by particles will be greater than that of the horizontal momentum flux of particles, which in turn will be greater than the height of the maximum mass flux of particles.

In the subsequent sections of this paper, however, the horizontal fluxes will be of the entire vertical column, which will include all moving particles.

Horizontal mass flux vs. deposit length

Horizontal mass flux, q, is defined as total mass flux collected during a unit duration of time through a surface that is perpendicular to the ground and to the wind and having unit width.

$$q = \frac{\int_0^T \int_0^w \int_0^H c\, V\, dz\, dy\, dt}{\int_0^T \int_0^w dy\, dt} \tag{7}$$

where H is height of the wind tunnel ceiling.

Values of q were divided by $\rho\, U^3/g$, which has the same units as q, and in which U is mean wind speed, ρ is density of air, and g is gravitational acceleration. The values $qg/\rho U^3$ plotted in Figure 7 show a definite increase with increase of deposit length ℓ. When q is divided by $\rho u_*^3/g$, however, and plotted against ℓ (Figure 8), the ratio is more-or-less constant with distance. That is, the increase of horizontal mass flux q with particle deposit length is explained by the feedback mechanism discussed by Owen and Gillette (in press) in their explanation of increased u_* with distance of particle deposit: the saltating particles accelerate causing a deceleration of the air layer in which the particles are embedded. This slowing of air near the surface promotes increased transport of momentum from above (increased stress), which in turn increases particle entrainment from the deposit. Effect of the finite ceiling height of the wind tunnel is also important in explaining the increase of u_*/U with ℓ. This effect is completely discussed by Owen and Gillette (1986).

Momentum fluxes

Horizontal momentum fluxes
Horizontal momentum flux of particles may be written:

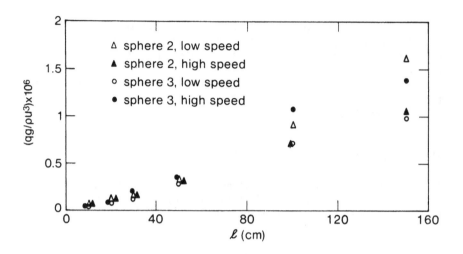

Figure 7 (Horizontal mass flux of particles)/$(\rho U^3/g)$ versus length ℓ from the leading edge of the eroding particle deposit.

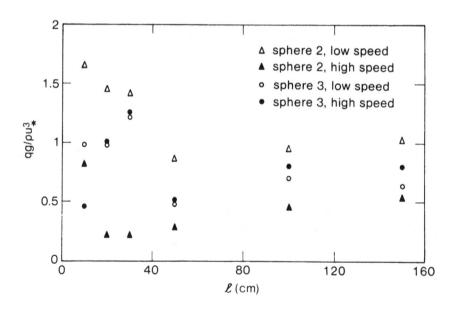

Figure 8 (Horizontal mass flux of particles)/$(\rho u_*^3/g)$ versus length ℓ from the leading edge of the eroding particle deposit.

$$\dot{M}_h = \frac{\int_0^T \int_0^w \int_0^H c\, V^2\, dz\, dy\, dt}{\int_0^T \int_0^w \quad dy\, dt} \qquad (8)$$

When divided by the quantity $\rho U^2 H$ which has the same dimensions (dynes cm^{-1}), and plotted against particle deposit length, ℓ , (see Fig. 9), $\dot{M}_h$ apparently increases with ℓ nonlinearly. When $\dot{M}_h$ is divided by q, however, a quantity having the dimensions of velocity results. This quantity, when divided by u$_*$ (given in Table 4), is a nondimensional ratio that does not show any clear trend for particle size, deposit length, or speed. The mean value with standard deviation of $\dot{M}_h/qu_*$) is 2.41 ± 0.95.

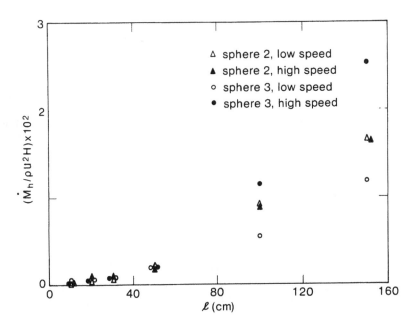

Figure 9 (Horizontal momentum flux of particles)/($\rho U^2 H$) versus length ℓ from the leading edge of the eroding particle deposit.

Table 4 $\dot{M}_h/(u*q)$ vs. ℓ

Sphere	Speed	Deposit length (cm)	$\dfrac{\dot{M}_h}{u*q}$
2	Low	150	3.03
		100	3.27
		50	2.32
		30	2.56
		20	1.89
		10	(0.47)
	High	150	2.36
		100	2.03
		50	0.92
		30	1.26
		20	2.01
		10	2.77
3	Low	150	3.05
		100	2.42
		50	2.26
		30	4.41
		20	4.51
		10	(9.82)
	High	150	3.15
		100	1.82
		50	1.07
		30	1.51
		20	1.95
		10	(2.91)
Mean			2.41
St. Dev.			0.95
N			21

Values in parentheses were doubtful and were not used in forming the mean.

Ratios of horizontal particle momentum flux with floor stress
The ratio of horizontal particle momentum flux with vertical momentum flux can be expressed:

$$\frac{\dot{M}_h}{\tau} \approx 1.8 \; \frac{u_*^2}{g} \tag{9}$$

Owen (1980) expressed the height of the saltation layer, H_s, to be proportional to u_*^2/g. Thus,

$$\frac{\dot{M}_h}{T} \propto H_s \qquad (10)$$

Particle kinetic energy fluxes

Horizontal
Although we did not measure kinetic energy fluxes directly, we can estimate them from the quantities previously reported. The horizontal kinetic energy flux is defined:

$$\dot{E}_h = \frac{1/2 \int_0^T \int_0^w \int_0^H c\, V^3\, dz\, dy\, dt}{\int_0^T \int_0^w dx\, dt} \qquad (11)$$

It is estimated as:

$$\dot{E}_h \approx 1/2\, \frac{\dot{M}_h^2}{q} \approx 1.2\, \dot{M}_h\, u_*. \qquad (12)$$

Vertical
For distances at which the surface stress is carried primarily by particle impacts, we can write:

$$\frac{d(\dot{E}_v)}{dx} = \frac{1/2 \tan\theta \int_0^T \int_0^w c(z=0)\, V^3\,(z=0)\, dy\, dt}{\int_0^T \int_0^w dy\, dt} \qquad (13)$$

where θ is the angle of the saltating particles to the ground at the end of their trajectories, or

$$\frac{d(\dot{E}_v)}{dx} = 1/2\, \rho\, (u_*^2 - u_{*_t}^2)\, \bar{v}(z=0) \qquad (14)$$

based on Owen's (1964) assumption that a portion of shear stress $\rho u_{*_t}^2$ is carried by fluid motion in well developed saltation. By considering the large increase of $\dot{M}_h$ with distance for the first 100 cm, the surfaces stress was carried dominantly by particle impaction of the surface only for distances greater than 100 cm. Actually, we did measure v(z=0-9 mm), not $\bar{v}$(z=0). Thus we use for the approximation:

$$\frac{d(\dot{E}_v)}{dx} = 1/2 \, \rho \, u_*^2 \, \bar{v}_{(z=0\text{-}9 \text{ mm})} \qquad (15)$$

We plotted the ratio of $\bar{v}$(z=0-9 mm)/u_* versus deposit length (Fig. 10). Lower values of u_* appear to correspond to larger ratios of v(z=0-9 mm)/u_*. We hypothesize that this is because saltation height grows as the square of u_*, and the height interval represents a larger fraction of particle layer depth for smaller u_*. Since we wish to approximate the value at the surface, it would seem that the values for higher speeds and lengths greater or equal to 50 cm would more appropriate to Equation 15. This average is v(z=0-9 mm)/u_* = 0.67 $\pm$ 0.14. This value is smaller than Owen's (1980) value for the velocity of a saltating particle at the end of its trajectory (about 6 u_*). This difference probably reflects that our average value is not a velocity at the end of the trajectory, but rather an average of particle speeds in differing parts of their saltation and creep trajectories. Thus, we would expect our number to be significantly smaller than the velocity at the end of a saltation trajectory.

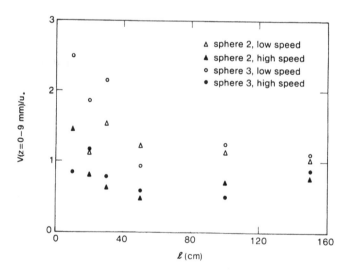

Figure 10 (Average velocity of particles moving at a height 0–9 mm above the surface)/(friction velocity) versus length ℓ from the leading edge of the eroding particle deposit.

Considering ratios of vertical kinetic energy by sandblasting to the surface to total horizontal mass flux, we may write for $\ell > 100$ cm:

$$\frac{d(E_v)/dx}{q} = 0.89 \, g \tag{16}$$

That is, the relationship of vertical sandblasting kinetic energy flux to horizontal mass flux becomes almost constant for particle deposit lengths greater than 100 cm and large friction velocities and is of the order 875 cm s^{-2}. The relationship of horizontal particle kinetic energy flux to vertical particle kinetic energy flux for distances greater than 100 cm from the leading edge of the deposit is:

$$\left[\frac{\dot{E}_h}{\frac{d(E_v)}{dx}} \right] = 6.5 \, u_*^2/g \tag{17}$$

That is, the ratio of horizontal to vertical kinetic energy flux from particles is roughly proportional to saltation layer height, and is quite large for large friction velocities.

Conclusions

The rather rapid changes of mass, momentum, and energy fluxes seen near the leading edge of an erodible particle deposit in a wind tunnel may be scaled by changes of friction velocity. The rapid change of friction velocity with distance in a wind tunnel has been explained by Owen and Gillette (in press) as due to a particle-air feedback mechanism that equivalently increases roughness height with distance. Horizontal mass, momentum, and kinetic energy fluxes may be scaled by friction velocity to the third, fourth, and fifth power respectively. Vertical momentum and kinetic energy fluxes for distances greater than 1 m from the leading edge may be scaled by friction velocities to the second and third powers respectively. Thus, the vertical kinetic energy flux and horizontal mass flux bear a constant ratio for distances greater than 1 m. For smaller distances, however, the ratio is smaller, decreasing to zero very near the leading edge of the erodible particle deposit.

Profiles of average particle speed and particle concentration showed that particle concentration decreases with height while average particle speed increases with height. Thus heights of maximum horizontal mass flux, momentum flux, and kinetic energy flux are successively larger.

Acknowledgements

The authors gratefully acknowledge Dr. John Adams of Riverside, California, for his enthusiastic support and encouragement of this project and Prof. P.R. Owen of the Imperial College of London for many extremely valuable discussions. The authors are also pleased to acknowledge Dr. Steven Clifford and Mr. Richard Lataitis of the NOAA Wave Propagation Laboratory for their help in comparing particle speed calculations with measurements of the calibration particle speed by their optical scintillation device. The National Center for Atmospheric Research kindly furnished wind measurement equipment and allowed us use of the anemometer calibration wind tunnel. Mr. D.W. Fryrear of the ARS/USDA furnished parts of the wind tunnel.

References

Bagnold, R.A. 1941. *The physics of blown sand and desert dunes*. London: Methuen.

Fuchs, N.A. 1964. *The mechanics of aerosols*. New York: Pergamon Press.

Gillette, D.A. 1978. Tests with a portable wind tunnel for determining wind erosion threshold velocities. *Atmospheric Environment* 12, 2309-13.

Greeley, R., S.H. Williams, B.R. White, J.B. Pollack, and J.R. Marshall 1983. Wind abrasion on Earth and Mars. *Models in geomorphology*. SUNY Buffalo: 14th Annual Geomorphology Symposium.

Hagen, L. 1984. Soil aggregate abrasion by impacting sand and soil particles. *Journal American Society of Agricultural Engineers*.

Horikawa, K. and H.W. Shen 1960. *Sand movement by wind action (on the characteristics of sand traps)*. Technical Memoir No. 119, U.S. Army Corps of Engineers.

Owen, P.R. 1964. Saltation of uniform grains in air. *Journal of Fluid Mechanics* 20, 225-42.

Owen, P.R. 1980. *The physics of sand movement*. Lecture notes for autumn course in The Physics of Flow in the Oceans, Atmosphere and Deserts. Trieste, Italy: International Centre for Theoretical Physics.

Owen, P.R. and D.A. Gillette (In press). Wind tunnel constraint on saltation. *Proceedings of the International Workshop on the Physics of Blown Sand*. Aarhus, Denmark, May, June 1985.

Schmidt, R.A. 1977. *A system that measures blowing snow*. USDA Forest Service Research Paper RM-194.

Wang, Ting-i, G. Lerfald, R.S. Lawrence and S.F. Clifford 1977. Measurement of rain parameters by optical scintillation. *Applied Optics* 16, 2236-41.

White, B.R. and J. Schulz 1977. Magnus effect in saltation. *Journal of Fluid Mechanics* 81, 497-512.

4

Particle transport by atmospheric winds on Venus: an experimental wind tunnel study

Bruce R. White

Abstract

High-speed motion pictures of saltating particles were taken in a high-pressure wind tunnel that dynamically simulates particle flow on the surface of the planet Venus. The high-pressure wind tunnel, called the Venus Wind Tunnel (VWT), reproduced the same dynamic pressure that saltating particles would experience under Venusian surface conditions. Velocity profiles taken within the VWT, displaying a mature turbulent boundary layer necessary for proper simulitude testing, are presented along with an analytically derived velocity profile for use in numerical solutions of particle motion. Analysis of experimental particle trajectories show the presence of a substantial particle spin force in the intermediate stages of the trajectories. The spin force occasionally acted (10-20% of the measured trajectories) in the direction of gravity producing a negative "lift". This phenomenon has not been previously observed at lower ambient pressures. Typical measured spin rates were in the order of several hundred revolutions per second and substantially influenced the particle trajectories. Numerical integration of the equations of motion including the spin force produced reasonable agreement with experiment.

Introduction

The Pioneer-Venus mission in the late 1970's, while probing the atmosphere of Venus, detected upper atmospheric wind speeds that suggest the existence of 1-2 m/s near surface wind speeds

(Counselman *et al.* 1979).   Earlier, in 1975, the Soviet landers
(Veneras 9 and 10) measured surface wind speeds of up to 1 m/s
(Florensky *et al.* 1977) at about a height of 1 m above the surface.
These surface wind speeds are within the range of predicted wind
speeds necessary to transport unsuspended solid surface material in
a saltating process (Greeley *et al.* 1984, Iversen & White 1982).

Typical saltation trajectories were calculated (White 1981) and
showed that the average saltation heights and lengths were much
smaller than those that would occur on Earth under dynamically
similar conditions (see Fig. 1).   The recognition of smaller scale
particle pathlengths may play an important role in interpreting
future radar images of the planet's surface.

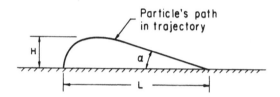

Figure 1 A typical path of a particle's trajectory. H is the maximum height the particle obtains in flight
while L is the maximum path length obtained; *a* is the collision angle which the particle's path makes
with the surface upon impact.

The exact nature of particle movement on the surface of Venus
is, however, not known.   There may be inaccuracies in the
numerical modelling of the saltation due to large-scale Reynolds
number effects on the particle motion.   In natural flow, there is
also a distribution of particle trajectories about the average.   The
range of these particle trajectory distributions is, of course, not
known from numerical modelling of the equations of particle
motion.

To more fully understand the particle motion process on Venus, a
wind tunnel was constructed to model the near-surface wind flow
on Venus. The high temperatures that exist on Venus are difficult
to simulate in the laboratory or in a wind tunnel specifically;
thus, at the expense of introducing some uncertainty, the wind
tunnel simulated dynamic, but not thermal conditions. Fortunately,
the thermal conditions are known not to play a significant role in
the physics of particle motion in the Venusian environment (White
1981).   Density is the principal atmospheric property governing
particle motion; thus, matching the Venusian density was the
primary impetus for constructing the high-pressure wind tunnel.
Matching density ratios will maintain dynamic similarity between
the wind tunnel and Venus.

This paper discusses the equations of motion of saltating particles as they apply to the high-pressure flow conditions which in turn apply to particle motion on the surface of Venus. The governing equations of motion are developed and include discussions on the wall-bounded velocity profiles. The experimental equations of motion are numerically solved for the same conditions as the experiments. Also presented are experimental data for the aeolian particle trajectories as photographed in the high-pressure wind tunnel. A comparison is made between the filmed trajectories and those that are numerically calculated.

Research facility and instrumentation

The experiments were conducted in a high-pressure wind tunnel that was especially built to simulate atmospheric flow on the surface of Venus. The high-pressure test facility is consequently named the Venus Wind Tunnel or VWT.

The VWT is a closed-circuit, boundary-layer tunnel with an overall dimension of 6 m by 2.3 m. Flow is generated by a one-horsepower, D.C. variable-speed motor which drives an eight-bladed fan. The fan operates at a maximum speed of 1750 rpm, which generates freestream wind speeds as high as 4 m/s through the test section (see Fig. 2).

Figure 2A Photograph of the high-pressure wind tunnel. It is a closed-circuit, atmospheric boundary-layer tunnel which operates with carbon dioxide or air over a range of pressure from atmospheric to 40 atmospheres of pressure.

The shell of the tunnel is constructed of schedule 40 steel pipe that has been hydrostatically tested to a pressure of 67 bars. Ten curved tubes, each 5 cm in diameter, are located in the two corners of the tunnel downwind from the fan to prevent flow separation and to minimize turbulence and secondary flow. A large diameter (47 cm) settling chamber containing Hexcel honeycomb and four 180-mesh screens is located immediately upstream of the test section to dampen small-scale turbulence. A smoothly-curved bell mouth provides the transition from the settling chamber to the test section. A diffuser section downstream from the test section improves the efficiency of flow through the tunnel. Located in the diffuser are electrical ports to provide 20 connectors for various instruments used in the tunnel. A replaceable screen between the diffuser and the fan assembly traps sand particles to prevent recirculation of the particles through the tunnel and damage to the fan.

The test section is 20 cm in diameter and 122 cm long, is mounted on a wheeled-track assembly, and has quick-couple clamps to enable easy access to the test section between experiments. Floor plates provide a level surface for experiments and are of several varieties depending upon the type of experiment. Four glass ports, each 7 cm in diameter, enable direct observation and photography of experiments in the test section.

The VWT operates with either carbon dioxide, nitrogen, or air, at pressures up to 40 bars. Although not duplicating the atmospheric temperature or pressure of the Venusian surface, by maintaining near-ambient (laboratory) temperatures the density of the Venusian atmosphere is duplicated at a pressure of 30 bars with CO_2. Although it would be desirable to duplicate the temperature and pressure on Venus, the cost of fabricating such a system is prohibitive. Thus, because the ratio of test particle-to-atmospheric density is the primary parameter of concern for most experiments dealing with the physics of windblown grains, the approach opted for was using gases at lower pressures and ambient temperature. Moreover, under VWT conditions the kinematic viscosity of the gas is approximately one half that on Venus, thus doubling the unit Reynolds number which is an advantage when simulating certain aspects of aeolian activity on Venus such as saltation and other processes in which Reynolds number is important.

Carbon dioxide is used for most of the experiments simulating Venus. It is stored as a liquid at 20 bars and 250 K and is brought to operating pressure (30 bars) and temperature (295 K) using a heat exchanger; it is then vented as a gas into the VWT. Nitrogen is also used in some experiments. It is supplied from a compressed gas storage chamber at 160 bars, then regulated to a lower pressure and vented into the VWT. In addition, air at 1 bar is used in the VWT for experiments to compare Venusian simulations with aeolian processes on Earth.

Gas temperature is monitored with a Leeds and Northrup Model

935 Numatron thermocouple system and type J (iron-constantine) thermocouples. The output signal is linearized with a model 939 auxiliary instrument and recorded on a two-channel strip chart recorder.

Saltation theory

The movement of an unsuspended solid over a gravity bed by wind action is known as saltation. The minimum friction speed at which the unsuspended solids begin to lift off the surface initiating saltation is known as the threshold friction speed. Bagnold (1941) has derived an expression for the threshold friction speed, u_{*t}, as a function of relevant parameters in the form of

$$u_{*t} = A \ (\rho_p g D_p / \rho)^{\frac{1}{2}} \qquad (1)$$

where ρ is the fluid density, ρ_p the particle density, D_p the particle diameter, and g the acceleration due to gravity. The dimensionless group A was assumed by Bagnold only to be a function of the friction-speed Reynolds number $R_f = u_* D_p / \nu$, where is the kinematic viscosity.

The coefficient A is nearly constant for values of R_f greater than about five; however, for values of R_f less than five, A increases rapidly with decreasing values of R_f. Iversen and White (1982) have presented new formulations valid for wide ranges of particle diameter including values of R_f substantially less than five for predicting values of A. These predictions, achieved by use of a low-pressure wind tunnel, allow one to separate the effects of Reynolds number and interparticle forces of cohesion. The formulations were used to predict saltation threshold for atmospheric conditions on Earth, Mars and Venus. The Venus threshold predictions, as well as those experimentally measured (Greeley et al. 1984), are used in the numerical simulation of particle trajectories presented.

Experimental velocity profiles

The measurement of the flow characteristics was accomplished by use of a calibration floor plate (smooth wall) that contained 22 static surface measurement ports and a boundary-layer Pitot-tube rake. The pressure ports from either the floor plate or the Pitot-tube rake were connected to a Model J Scanivalve system. The Scanivalve system was connected to a Barocel pressure transducer which provided a pressure signal that was recorded for later data reduction. Both the transducer and Scanivalve system were located downstream of the test section in the corner of the wind tunnel diffuser section to minimize interference with flow through the test section.

Mean-flow profiles were determined from the differential pressure measurements of the boundary-layer rake. They were reduced from strip chart recordings of the pressure differential as a function of height. The profiles were required to have constant ambient tunnel pressure ($\pm 0.5\%$) and a constant freestream velocity ($\pm 2\%$). Data were then recorded from the mean profile. Each profile data set contained at least 20 readings, the majority having many more than 20 data points.

Each profile is numerically curve-fitted by means of a multipiece cubic spline technique. The displacement thickness δ^* and momentum-deficit thickness θ are determined by integrating the resultant curve fit using a numerical quadrature technique.

The surface shear stress τ_o can be determined by Cole's (1962) method from the velocity profiles. This is accomplished by assuming the velocity profile obeys the logarithmic law

$$\frac{u}{u_*} = \frac{1}{K}\log \frac{u_* y}{\nu} + C \qquad (2)$$

for the wall region ($y/\delta < 0.2$) excluding the viscous sublayer where δ is the boundary layer height. This technique is utilized in the present study to determine τ_o using $K = 0.418$ and $C = 5.45$ as determined by Patel (1965). Hence,

$$u^+ = 5.5 \log_{10} y^+ + 5.45$$

Figures 3 and 4 show u/u_∞ as a function of y/δ and u^+ as a function of Reynolds number, $u_\infty y/\nu$ for a typical velocity profile measured in the VWT. The C_f determined from the Clauser (1954) technique is 0.00210 for Figure 3. This compares well with a value of 0.00203 which was calculated from Coles's (1968) empirical expression,

$$C_f = \frac{0.3 \exp(-1.33H)}{(\log R_\theta)^{1.74 + 0.31H}} \qquad (3)$$

where H is the shape factor defined as δ^*/θ. The Reynolds number, based upon freestream speed and downstream distance, was 14.5 million and 8.81 million, respectively, for Figures 3 and 4.

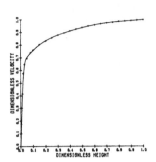

Figure 3 u/u$_\infty$ as a function of y/δ and the cubic-spline curve fit (solid line). Test conditions: u/$_\infty$415 cm/ s, VWT pressure equals 30.6 atmospheres, kinematic viscosity equals 0.00232cm^2/s and the Reynolds number equals 14.5 million where x is the downstream fetch; δ = 2.06 cm.

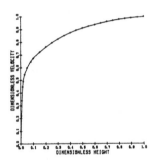

Figure 4 u/u$_\infty$ as a function of y/δ and the cubic-spline curve fit (solid line). Test conditions: u$_\infty$ = 248 cm/s, VWT pressure equals 35.7 atmospheres, kinematic viscosity equals 0.00212 cm^2/s and the downstream Reynolds number equals 8.81 million; δ = 2.85 cm.

Analytical velocity profiles

A one-dimensional flow situation may be assumed in which the velocity in the vertical (y) direction is zero and the velocity u in the flow (x) direction is a function of height about the surface.

For atmospheric pressure flows where the surface is aerodynamically rough, the velocity profile is modelled by the equation

$$\frac{u}{u_*} = 5.5 \log_{10} (y/y_0) \tag{4}$$

where y_o is the roughness height and may be taken equal to $D_p/30$ for a surface of like particles. Boundary layer flows at Venus surface pressures are usually fully turbulent and have no viscous sublayer.

Particle motion significantly alters velocity profiles. Bagnold (1941 and others) demonstrated experimentally that the velocity profile may be expressed by

$$\frac{u}{u_*} = 2.5 \ln (y/y_o') + u_o'/u_*. \tag{5}$$

Here y_o' is the roughness height for saltating grains as defined by Bagnold, and u_o' is an average speed of the top layer of surface particles.

When threshold speeds are attained there are few particles in the air and the boundary layer tends to retain approximately the profile of non-particle-motion flow with a viscous sublayer beneath it. Hence,

$$u_o' / u_{*_t} = 2.5 \ln R_f + 6.74 \tag{6}$$

where R_f is the friction Reynolds number, $u_* D_p/\nu$.

The height y_o is a function primarily of D_p and, on the basis of Andres' (1970) experimental data, y_o' is replaced by $2.5 D_p$. Substitution of this relationship and of Equation 6 into Equation 5 yields the final form of the velocity-profile equations at the initiation of saltation as

$$u = 2.5 u_* \ln (y/D_p) - 2.29 u_* + u_{*_t} (2.5 \ln R_f + 6.74) \tag{7}$$

Principles of particle motion

A one-dimensional flow situation is assumed to calculate the trajectories in which the velocity in the vertical (y) direction is zero and the velocity u in the flow (x) direction is a function of height above the surface only (see Fig. 5). When the saltation level is low, Equation 7 may be used to describe the velocity profile and it was used for the following numerical calculations.

The forces acting on a saltating particle are a downward force due to gravity, the aerodynamic drag D acting opposite the direction of relative particle velocity V_r, and the lift L acting

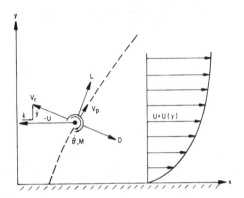

Figure 5 Forces and velocities associated with a saltating particle.

normal to the drag. The equations for translational motion of a particle can be written as

$$m_p \ddot{x} = L \frac{\dot{y}}{V_r} - D \frac{\dot{x} - u}{V_r} \tag{8}$$

$$m_p \ddot{y} = -L \frac{\dot{x} - u}{V_r} - D \frac{\dot{y}}{V_r} - m_p g \tag{9}$$

where m_p is the particle's mass, $(\dot{x},\dot{y})$ and $(\ddot{x},\ddot{y})$ are the particle's velocity and acceleration components, and g is the acceleration due to gravity.

In Equations 8 and 9,

$$V_r = [(\dot{x} - u)^2 + \dot{y}^2]^{\frac{1}{2}} \cdot \tag{10}$$

The effects of lift and drag are expressed in terms of lift and drag coefficients C_L and C_D, defined by

$$L = \frac{1}{8} C_L \pi \rho D_p^2 V_r^2 \quad \text{and} \quad D = \frac{1}{8} C_D \pi \rho D_p^2 V_r^2 \tag{11}$$

Assuming the particles are spherical and of uniform density, the equations of motion reduce to

$$\ddot{x} = -\tfrac{3}{4} \frac{\rho}{\rho_p} \frac{V_r}{D_p} \ [C_D (\dot{x} - u) - C_L \dot{y} \ ] \qquad (12)$$

$$\ddot{y} = -\tfrac{3}{4} \frac{\rho}{\rho_p} \frac{V_r}{D_p} [C_D \dot{y} + C_L (\dot{x} - u)] - g \qquad (13)$$

The drag coefficient of a sphere is strongly dependent upon the Reynolds number. The empirical formulas of Morsi and Alexander (1972) are used for its determination in the numerical calculations.

The substantial lift of a particle away from the surface is negligible except for that lift that is produced by the spinning of the particle. Particle spin (or Magnus effect) during saltation at atmospheric pressure has been shown to have a profound influence on the overall resulting trajectories (White & Schulz 1977, White 1982). Spin rates have been verified to be on the order of several hundred revolutions per second and possibly thousands of revolutions per second for high-speed compressible flow.

Rubinow and Keller (1961) derived the following relation for the lifting Magnus force on a rotating spherical particle:

$$L = \tfrac{1}{8} \pi \ D_p^3 \ \rho \ V_r \ (\dot{\theta} - \tfrac{1}{2} \frac{\partial U}{\partial x}) \ , \qquad (14)$$

$$M = \pi \mu \ D_p^3 \ (\tfrac{1}{2} \frac{\partial U}{\partial x} - \dot{\theta}) \ . \qquad (15)$$

where M is the moment of the particle. Equations 14 and 15 assume that the particle moment and lift are developed from the rotation of the particle. Here, $\dot{\theta}$ is the particle's angular velocity and μ the fluid viscosity.

Equations 14 and 15 are valid only for the case of extremely small Reynolds number. Their use in the present case is justified by the fact that it is desired primarily to determine whether the addition of terms accounting for the particle spin will significantly improve the agreement between theory and experiment.

Comparison of filmed and calculated trajectories

The high-speed motion picture films of saltating flows evaluated in this study were taken with a HiCam camera in the VWT wind tunnel with a test section speed of 1 to 4 m/s. The camera was pointed horizontally and perpendicular to the direction of flow such that the flow and the saltating particles travelled from left to right across successive frames. The flow was lit from above by a thin strip of high intensity light so that only relatively few particles in the direction transverse to the flow direction were illuminated, and hence visible on the film. A centimeter grid placed on the far side of the flow provided a spatial reference frame which permitted subsequent determination of the position of individual particles.

The saltating material was quartz with a density of 2.58 g/cm^3. Three different types of quartz, each with a different diameter range, were used, and each of these types was photographed at several different flow velocities. The films were taken at a nominal frame speed of 1024 frames per second. A strobe light device left a light flash on the margin of the film which permitted for a more precise determination of the film speed to be made later.

Before each filmed run, a layer of quartz was spread on the wind-tunnel floor, carefully smoothed out and levelled to a height of about 0.2 cm. The patch covered was approximately 10 cm by 122 cm, the longer dimension being in the direction of the flow. The flow friction velocity and freestream velocity during the runs were determined by measuring the pressure differential between the test section and an upstream position in the tunnel, the relationship between these velocities and the pressure differentials having been established previously from velocity traverses made with no "sand" in the tunnel.

In order to obtain trajectories of individual saltating particles, the film was analyzed using a Vanguard Motion Analyzer. The film could be run through this instrument forwards or backwards at either a continuously variable speed or frame by frame. A frame counter made it possible to keep track of position within the section of film being viewed. The film was projected onto a flat frosted-glass surface, the size of the projected image on this surface being about 5 cm by 9 cm. Crank-operated horizontal and vertical cross-hairs could be centered on any point on the screen, and its position read from dials to within one-thousandth of an inch (25.4 μm).

A number of trajectories were obtained from the film. These were compared with trajectories obtained by numerical integration of the equations of motion including only the drag (no lift) with starting values for the integration chosen as the initial position and velocity of the particles on the filmed trajectories.

Figure 6 displays a typical comparison between the filmed path traced out by a particle with two theoretical solutions. The first solution is with the experimentally measured spin rate of 64 revolutions per second. The plane of spin appears to be entirely in the vertical longitudinal plane as is the particle motion. Reasonable agreement is found between the filmed trajectory and the numerical solution with particle spin. The measured and calculated particle speeds along the trajectory were nearly the same for the solution with particle spin. The drag only comparison had poor agreement.

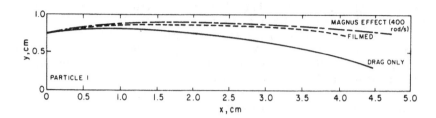

Figure 6 A comparison of a filmed (dashed line) path traced out by a particle with numerical calculations from the equations of motion with (dashed-dotted line) and without (solid line) measured particle spin of 64 rev/s (400 rad/s). The particle was approximately 550 microns. Test conditions are presented in Table 1.

In many other trajectories, the particle spin is not entirely in the vertical longitudinal plane as is the particle motion. This effect is perhaps due to the particle colliding with surrounding surface particles as it lifts off or rebounds from the surface. Figure 7 displays a filmed trajectory where the spinning of the particle is estimated from the film data which shows the particle spin not entirely located in the vertical longitudinal plane. The component of spin that is estimated to lie in the vertical longitudinal plane is 90 revolutions per second; good agreement is observed between the filmed trajectory and the numerical solution with the estimated vertical longitudinal component of particle spin. Abbott and Francis (1977) observed a similar phenomenon for saltating solid particle motion in water.

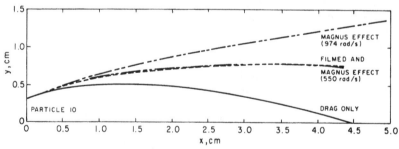

Figure 7 A comparison of filmed (dashed line) path traced out by a particle with numerical calculations from the equations of motion with (dashed-dotted-dotted line) and without (solid line) measured particle spin of 155 rev/s (974 rad/s). Here, the true spin plane was not the vertical longitudinal; spin was 90 rev/s (550 rad/s) shown by the dashed-dotted line. The particle was approximately 550 microns. Test conditions are presented in Table 1.

An interesting observation of some of the filmed trajectories, apparently a result of surface particle collisions and the highly inviscid high-pressure gas flow, is that many particles have negative spin rates thus producing a negative "lift" throughout the entire pathlength of the particle trajectory. Figure 8 displays a typical comparison between such a particle's trajectory with the drag only numerical solution and that of the experimentally measured spin rate of -239 revolutions per second. Here, the vertical longitudinal plane appears to lie entirely in the plane of particle spin, although negatively. This phenomenon occurs in approximately 10 to 20% of all saltating particles observed.

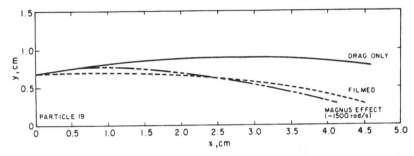

Figure 8 A comparison of a filmed (dashed line) path traced out by a particle with numerical calculations from the equations of motion with (dashed-dotted line) and without (solid line) measured particle spin of -239 rev/s (-1500 rad/s). The particle was approximately 550 microns. Test conditions are present in Table 1.

Comparison of filmed and theoretical trajectories

An entirely analytical saltation model (that considers only spherical particles in the numerical solution of particle trajectories) was previously developed (White *et al.* 1976) and applied to the case of Venus (White 1981). The theoretical saltation model needs no experimental data in order to solve the particles' trajectories. The calculation procedure outlined in the previous section does require experimentally specified initial conditions in order to generate trajectory solutions. For the theoretical trajectory model, a particle is at rest on the surface, and as the wind speed is increased above threshold, particles begin to move.

A good check on the theoretical calculations is to compare the experimental values of maximum height H and maximum jump length or pathlength L (see Fig. 1) to the theoretically obtained values. In a mature saltation process, one expects a range of particle trajectories due to inherently stochastic processing including a range of H's and L's for a given particle size. However, the average values of H and L should agree reasonably well with the theoretically calculated values. Figures 9 and 10 show the dependence of the "typical" maximum height H and pathlength L of saltating particles on the particle diameter D_p for identical conditions to those present in the VWT experiments given in Table 1. For a D_p of 550 microns, the H value is 0.99 cm while the pathlength L is 8.32 cm. Of twenty-six trajectories observed in the VWT under identical conditions as those in the theoretical trajectory calculations, the average H value was 0.97 cm with a standard deviation of 0.199. Although the experimental pathlength could not be measured directly (due to limitations of the movie camera's field of view) the extrapolated pathlengths appear to be close to the theoretically determined value of L = 8.32 cm. The agreement is remarkable and demonstrates the apparent accuracy of the theoretical saltation model.

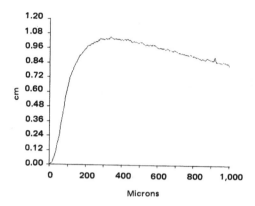

Figure 9 Maximum vertical heights H for saltating particles as a function of particle diameter D_p. This is from results of theoretically calculated trajectories with conditions presented in Table 1.

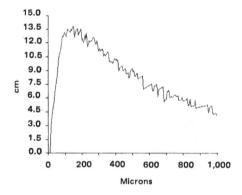

Figure 10 Maximum path lengths L for saltating particles as a function of particle diameter D_p. This is from results of theoretically calculated trajectories with condition presented in Table 1.

Figure 11 displays a plot of contours of constant particle pathlengths as functions of wind speed and VWT chamber pressure for particle diameters of 100 microns. The same analytical method of calculation utilized by White (1981) was employed for the series of calculations that generated the contour map. As observed, as the chamber pressure is increased, the particle pathlength becomes independent of chamber pressure. In fact, at Venus surface densities (pressure equal to about 30 bars), the particle pathlength is almost linearly related to freestream wind speed. An interesting

Table 1 Values of the parameters used in the saltation calculations.

Parameter	Symbol	Value
surface pressure	p	3.068 MPa (455 psi)
gas temperature	T	22 C
acceleration due to gravity	g	980.7 cm/s
Von Karman constant	K	0.41
atmospeheric gas composition	—	carbon dioxide
particle density	ρ_p	2.58 g/cm³
particle diameter	D_p	0.055 cm
wind speed	u	233 cm/s
fluid density	ρ	0.0368 g/cm³
friction velocity	u_*	10.610 cm/s
fluid viscosity	ν	0.0057 cm²/s
threshold friction velocity	u_{*t}	17.42 cm/s

feature of Figure 11 is that the lower contour of constant particle pathlength of two centimeters approximately coincides with the values of particle thresholds from VWT experiments (Greeley *et al.* 1984). Evidently, the threshold condition is associated with recognizable particle motion of a critical minimum dimension. In the case of the VWT, it appears as though that recognizable motion is a particle pathlength of roughly two centimeters. Close examination of the high-speed motion films reveals that even the rolling mode of particle transport consists of minute particle leaps, and in fact, no particle is ever exclusively observed to roll across the surface. At normal viewing speeds, these minute saltation leaps appear to make the particle "roll" along the surface while actually the particles are saltating on a miniature scale.

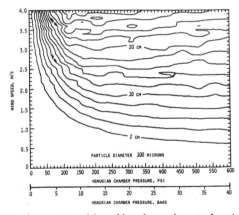

Figure 11 Contour lines of constant particle pathlengths are shown as functions of freestream wind speed in the VWT and ambient VWT chamber pressure. These results are produced from the theoretical saltation model for a particle size of 100 microns.

Summary

A high-pressure Venus Wind Tunnel (VWT) was constructed in order to investigate the behavior of windblown particles in a simulated Venusian environment. The flow through the VWT was carefully documented and was found to have good near surface boundary-layer profiles for simulating particle motion under Venusian surface conditions. Experimental surface shear stress values were determined and subsequently used in numerical calculation of particle motion. In addition, an analytical velocity profile was derived for use in the fully theoretical calculation of particle motion.

Experimental measurements of particle trajectories were made that dynamically simulated particle motion on the surface of Venus. Numerical integration of the equations of motion using experimentally determined initial conditions including a particle spin force term produced reasonable agreement with the experiment. Typical spin rates were in the order of several hundred revolutions per second, and the transverse force produced from the spinning of the particles had a profound influence on the nature of the particles' trajectories. Some 10 to 20% of particles observed had negative spin which produced negative "lift" tending to shorten the pathlength and maximum height of the trajectories. This phenomenon has not been previously observed in saltating flow.

Comparison of experimental particle trajectories were made with the fully theoretical calculations using the saltation model described by White (1981). There was good agreement between the theoretical predictions and the observed average of experimental data for the limited cases presented. The agreement suggests the saltation model may be used with confidence in calculating particle motion on the surface of Venus.

Acknowledgement

This research was supported by the office of Planetary Geology, National Aeronautics and Space Administration, through interchange Agreement NCA2-OR180-903 to the University of California at Davis. The author wishes to express his sincere graditude to Ms. Linda Smith who performed much of the data reduction of the VWT particle trajectories.

References

Abbott, J.E. and J.R.D. Francis 1977. Saltation and suspension trajectories of solid grains in a water stream. *Philosophical Transactions of the Royal Society of London* 284: 1321, 225-54.

Andres, R.M. 1970. *The mechanics of dusting lifting with particular emphasis on the planet Mars.* Ph.D. dissertation, St. Louis University.

Bagnold, R.A. 1941. The physics of blown sand and desert dunes. London: Methuen.

Clauser, F.H. 1954. Turbulent boundary layers in adverse pressure gradients. *Journal of Aeronautical Sciences* 21, 91-108.

Coles, D.E. 1962. *The turbulent boundary layer in a compressible fluid.* Rand report 403-PR.

Coles, D.E. and E.A. Hirst (eds.) 1968. *Proceedings, computation of turbulent boundary layers.* Volume 2 AFOSR-IFP, Stanford Conference.

Counselman, C.C., S.A. Gourevitch, R.W. King, G.B. Loriot and R.G. Prinn 1979. Venus winds are zonal and retrograde below the clouds. *Science* 205, 85-7.

Florensky, C.P., L.B. Ronca, A.T. Basilevsky, G.A. Burba, O.V. Nikolaeva, A.A. Pronin, A.M. Trakhtman, V.P. Volkov and V.V. Zazetsky 1977. The surface of Venus as revealed by Soviet Venura 9 and 10. *Geological Society of America Bulletin* 88 (11), 1537-47.

Greeley,R., J.D. Iversen, R. Leach, J. Marshall, B. White and S. Williams 1984. Windblown sand on Venus: preliminary results of laboratory simulations. *Icarus* 57, 112-24.

Iversen, J.D. and B.R. White 1982. Saltation threshold on Earth, Mars and Venus. *Sedimentology* 29, 111-19.

Morsi, S.A. and A.J. Alexander 1972. An investigation of particle trajectories in two-phase flow systems. *Journal of Fluid Mechanics* 55, 193-208.

Patel, V.C. 1965. Calibration of the Preston tube and limitations on its use in pressure gradients. *Journal of Fluid Mechanics* 23, 185-208.

Rubinow, S. and J. Keller 1961. The transverse force on a spinning sphere moving in a viscous fluid. *Journal of Fluid Mechanics* 11, 447-59.

White, B.R. 1981. Venusian saltation. *Icarus* 46, 226-32.

White, B.R. 1982. Two-phase measurements of saltating turbulent boundary layer flow. *International Journal of Multiphase Flow* 8 (5), 459-73.

White, B.R. and J.C. Schulz 1977. Magnus effect in saltation. *Journal of Fluid Mechanics* 81, 497-512.

White, B.R., R. Greeley, J.D. Iversen and J.B. Pollack 1976. Estimated grain saltation in a Martian atmosphere. *Journal of Geophysical Research* 81 (32), 5643-50.

5

The role of vegetation in the formation of linear sand dunes

Haim Tsoar and Jens Tyge Møller

Abstract

A recent change in the vegetation cover of linear dunes in the Negev desert forms the basis for the study of the effect of vegetation on linear dune morphology. Linear dunes are elongated with little erosion of sand. The upshot is more protection of sand moisture which in turn sustains the relatively dense plant cover on this particular dune type in deserts. Destruction of the vegetation brings about abrupt change in the morphology of linear dunes and turns them into linear braided and seif dunes. This process was analyzed by aerial photographs and computer-aided photogrammetric mapping. Vegetated linear dunes align parallel to dominant wind direction while seif dunes stretch between two main dominant wind directions blowing on both sides of them at an angle of $30° \pm 10°$ with the dune crestline.

Introduction

Several types of linear dunes are identified in the world's deserts. It is commmon to divide them and other dune types into simple, compound and complex forms (Melton 1940, Holm 1968, McKee 1979, Lancaster 1982). Compound and complex linear dunes are, in most cases, megadunes devoid of vegetation and are found in the driest part of the deserts (Breed *et al.* 1979). Simple linear dunes are known from all parts of the world's arid, semi-arid and humid lands; they are sometimes covered, partially or wholly, by vegetation or have none at all. Linear dunes cover approximately 30% of all aeolian depositional surfaces and represent 50%-65% of all dune types (Fryberger & Goudie 1981, Lancaster 1982). Most of them are of the simple and relatively low type (Breed *et al.* 1979).

The advent of satellite images during the 1970's paved the way for an intensive analysis of dune forms of the world's deserts by the remote sensing approach (McKee *et al.* 1973, El-Baz 1978, 1984, Breed *et al.* 1979, Blom & Elachi 1981). This method allows for a broad worldwide survey of desert sand seas, but has the disadvantage of low resolution, since the morphology of the sand dunes and the variation in dune vegetation are not always perceptible. It is therefore customary to refer to the several types of parallel straight dunes by the general denomination of *linear dunes* (Breed *et al.* 1979, Fryberger & Goudie 1981, Lancaster 1982).

Area of research and its dune types

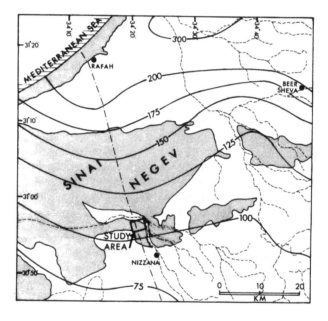

Figure 1 The area of research in the northwestern Negev and the average annual precipitation (in mm). Stippled areas indicate sand dunes. The enclosure shows the study area of Sede Hallamish.

The northwestern Negev and northeastern Sinai is factually one geographic unit divided artificially by a political border line determined in 1906 (Fig. 1). This area is located in a transitional zone between a semi-arid area in the north with an average of >200 mm annual rainfall and an extremely arid area in the south with no more than 50 mm. For this reason, one should not be surprised to find both vegetated and unvegetated dunes in the in-between area. There is, however, a comparatively sharp transition in vegetation between Negev and Sinai. The Negev linear dunes are partially vegetated and thereby similar to the

Australian sand ridges (Striem 1954). On the other hand, most of the northern Sinai linear dunes are devoid of the vegetation that is only found in the interdune area, thus resulting in typical sinuous seif dunes. This sharp contrast in vegetation is reflected very well on satellite images of the area, showing a prominent increase in surface albedo in Sinai in comparison with the Negev (Fig. 2). This phenomenon has attracted world-wide attention as a typical example of regional elimination of vegetation through overgrazing (Lowman 1966, Otterman 1974, 1981, Otterman *et al.* 1975, Schneider 1976, Muehlberger & Wilmarth 1977, Adams *et al.* 1978, Noy-Meir & Seligman 1979, Danin *et al.* 1982, Warren & Harrison 1984).

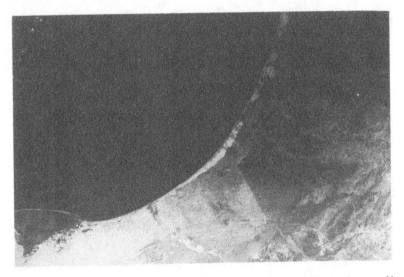

Figure 2 The research area on a LANDSAT image (MSS4) of December 1972. The northwestern Negev has a lower albedo than the northeastern Sinai because it is much richer in vegetation over the sand dunes.

The study area where the effect of vegetation on linear dunes was monitored and analyzed, is located in Sede Hallamish, in the southern part of the dune field, north of Nizzana (Fig. 1). The average annual rainfall in Nizzana is 78 mm and the relative standard deviation of rain is 68%. The highest record of rainfall there was 284 mm and the lowest 23.4 mm. The rainy season starts in October and ends in May, while most of the precipitation takes place between December and February.

Three types of simple linear dunes are distinguished in the research area: smoothly-rounded straight linear dunes, usually symmetrical in cross section yet sometimes asymmetrical, vegetated with uneven crests with more mobile sand, and a scantier vegetation cover (Figs 3 & 4). This type is known in Australia as *sand ridge* (Madigan 1946, Price 1950, Twidale 1972, 1981, Breed &

Figure 3 Vegetated linear dunes in the northwestern Negev.

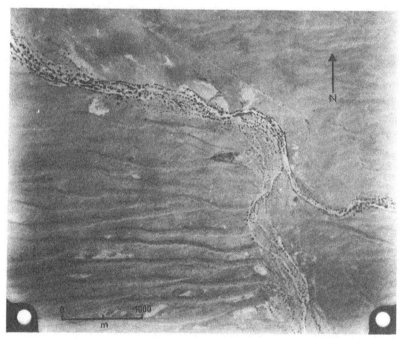

Figure 4 Aerial photograph of the study area taken on 26 July 1968, showing linear vegetated dunes with Y-junction.

Breed 1979) or as *parallel ridge* (Mabbutt 1968); it will be referred to here as *vegetated linear dune*. Like other similar types, they tend to converge, forming a Y-junction or a tuning fork shape (Folk 1971, Twidale 1972). Simple vegetated dunes cover most of the Australian deserts (Jennings 1968), southern Mauritania (Breed *et al.* 1979), the Kalahari region of southern Africa (Lancaster 1981) and some small areas in the southwestern United States (Breed & Breed 1979). There is a second type similar to the above in its straight lineament, yet different because of its much sparser vegetation, especially around the crest. Secondary transverse dunelets are superimposed on this type, thus forming a crested dune (Fig. 5) which is always asymmetrical in cross section. This type is referred to in this study as *linear braided* and classified as being a complex type (Mckee 1979). The third type, a sinuous, sharp-crested, generally symmetric one in cross section, is unvegetated and known mainly in the Sahara and Arabia as the *seif* dune (Bagnold 1941, 1951, Smith 1963, McKee & Tibbitts 1964, Warren 1972, Tsoar 1983). The term "seif" (sword in Arabic) indicates the crest's knife-edged shape (Fig. 5).

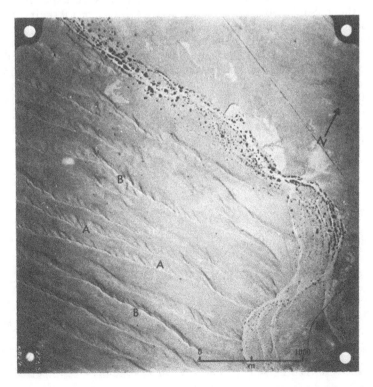

Figure 5 Aerial photograph of the study area taken on 24 April 1976. Two types of linear dunes are discerned: braided (A) and seif (B).

Recent changes in the vegetation cover of the research area

Sand dunes always exercised attraction on the Bedouin nomads of Sinai and Negev. The dune belt is located at the northern part of the area, where precipitation is higher (Fig. 1). The Bedouin uses the sand surface for growing castor, bean, barley, wheat and water melons which also fodder the camels during the dry summer season (Bailey, personal communication 1983).

An analysis of the aerial photographic survey taken along the Sinai-Negev line during 1944 and 1945 clearly shows that there was no contrast in albedo between the two areas and that the linear dunes, on both sides, were then covered by vegetation (Noy-Meir & Seligman 1979). The establishment of the State of Israel in 1948 brought about a radical change in the ratio of the Bedouin population in the Negev. In 1948 there were approximately 65 000 Bedouin in the Negev, but when the Israeli Army occupied the Negev at the end of 1948, many of them fled, and others were either relocated or expelled. In 1953 about 11 000 Bedouin remained and were concentrated by the Military Administration in the northeastern corner of the Negev, beyond the sand dune belt (Marx 1967). Part of the Bedouin that left or were expelled moved to northern Sinai. Because of the sudden increase in their number, the Bedouin in the sand dune area of northern Sinai led a wretched existence, which forced some of them to move to Jordan between 1960 and 1967 (Bailey & Peled 1974). As a result of an effectively closed border and intensified shrub removal on the Sinai side of the border, mostly for firewood and shelters (Zohar *et al.* 1971) and also for cultivation and livestock grazing (Bailey & Danin 1981), a sharp line of contrast between a "lighter" Sinai and a "darker" Negev was created, as thus observed in aerial and satellite photographs since the early 1960s (Fig. 2).

Figure 6 A vegetated linear dune of the study area that was overgrazed by flocks coming from Sinai (seen in the foreground). This dune is in the process of turning into linear braided type.

As a corollary to the war in 1967, the border line between Sinai and the Negev was opened up again until 1982, when it finally became the international border between Israel and Egypt. During this period, some of the Sinai Bedouin nomads exploited the open border and infiltrated the Negev (Fig. 6), with resultant severe overgrazing and vegetative destruction in some areas along the line. Two maps of the linear sand dunes of Sede Hallamish, covering an area of 1250 by 2500 m (3.125 km^2), were plotted by photogrammetric methods. One is based on aerial photographs taken in July 1968, and the other on photographs from April 1982. The map altitude accuracy of the aerial photographs is about ±0.5 m. The mapping contour data were fed to a computer which plotted 3-dimensional diagrams and calculated the ratio of sand change accumulation during the in-between period. The results are the basis for analyzing the transformation of linear vegetated dunes into linear braided and seif-type ones.

Dune sand as a substratum for desert vegetation

Active sand seas in the world's deserts occur mostly within the 150 mm precipitation line where vegetation is sparse or absent and, as a result, unimpeded sand transport can take place when the wind is strong enough (Wilson 1973). It seems obvious that most research papers on desert dunes did not include vegetation as a determinative factor in sand dune morphology and dynamics, even when Australia's linear vegetated dunes were dealt with (Folk 1971, Wilson 1972). Only in works on coastal dunes was the role of vegetation taken into account (Olson 1958, Pye 1982a, Goldsmith 1985). Twidale (1981) regards the vegetation cover in the Australian deserts as anomalous and assumes that it is a relic feature dating back to the Late Pleistocene.

In humid areas, dune sand is generally considered as barren inert soil with relatively poor vegetative vitality. However, reports by early travelers in the world's deserts indicated, with some astonishment, that vegetation on sand dunes in a dry country thrives more easily than elsewhere (Gabriel 1938). In the Negev desert, sand dunes support a sometimes perennial vegetation that is denser than that of heavy soils such as loess. In the northwestern Sinai, perennial vegetation on sand dunes is found in areas having 50 mm average annual rainfall and even less (Danin 1983). This apparent contradiction springs from the property of the sand's moisture retention as can be shown by the equation of the soil's water balance:

$$\Delta S = P\text{-}R\text{-}D\text{-}E \qquad (1)$$

where: ΔS = net change in soil moisture reserve

   P = total precipitation

   R = runoff and other horizontal flows

   D = infiltration to deeper soil layer

   E = evapotranspiration

The after rain soil moisture reserve depends on the soil's texture as determined by moisture tension brought about by the adhesive forces that bind water to soil matrix. The moisture tension of sandy soil is generally very much inferior to that of finer soils composed of silt and clay for the same amount of moisture in ratio. Therefore, in sand the infiltration to deep soil layer (D) is higher. The result is a relatively low availability of soil moisture for cultivation in dune sand.

However, in arid and semi-arid areas the situation is different. There the amount of precipitation (P) is very small, and as a result, D is negligible. Likewise, runoff (R) does not take place on dune sand, so that evaporation (E) from upper layers constitutes the main cause of soil moisture loss.   Because of its thermal properties, desert soils have a high propensity to dry up.   The capacity of water for deep-percolation depends on the soil texture, with sand having the deepest pervasion and relatively lower loss through evaporation (E). The water balance equation demonstrates that desert sand retains more water available for cultivation than finer soils composed of silt and clay. The equilibrium between the advantage of sand texture (low E) and its disadvantage (high D) stands at around 300 mm of average annual rainfall (Noy-Meir 1973).

As sand dunes with vegetative cover have been observed in areas with less than 100 mm rainfall, it is implicit that sand dunes in arid areas should be covered, as a natural process, by vegetation. Only dunes in extremely arid or hyperarid areas with almost no rainfall will be completely devoid of vegetative cover. In cases where no vegetation on sand dunes is found in arid or semi-arid areas, one should seek the reasons in the history of the area's use. Sand responds to the growing of vegetation in deserts in favourable relation to its surroundings; this contradicts the allegation that sand dunes develop mainly in deserts due to the latter's lack of vegetation. It is more appropriate to affirm that sand dunes form and exist in any place where sand is plentifully available and has a propensity to accumulate.

Control of vegetation on sand dune accumulation and development

Few works have emphasized the importance of vegetation in the evolution and development of sand dunes. It is accepted that vegetation creates some kind of surface roughness which traps sand (Bagnold 1941, Olson 1958, Pye 1982a, Hesp 1981). Vegetation initiates and regulates dune growth. Sand around the vegetation continues to accumulate as long as the plants grow higher. Pye (1982a) mentions two other ways in which vegetation retards the progress of sand movement. One is the binding effect of the root on the surface sand layers and the second is the increase of the cohesion and water-retentive properties of the surface sand by the humus derived from dead vegetation.

In deserts, vegetation acts also as a trap for particles suspended in the atmosphere, mainly of silt and clay. It is known that active desert dunes contain no more than 1%-3% of silt and clay. Any additional amount is taken back to suspension during sand storms through the impact of saltating grains (Tsoar 1976, Pye 1982b).

Vegetation reduces saltation and thereby contributes to increasing the dune's silt and clay content (Kutiel et al. 1979/80, Danin & Yaalon 1982). Figure 7 shows the amount of and variation in fines (< 0.062 mm) in three pits dug on vegetated semi-stabilized linear dunes. The highest amount of fines (up to 10%) is concentrated on the ground surface where suspended motes settle and form a silty crust. Rainwash leaches part of the particles to a depth of 30-50 cm which is the average penetration depth of rain in deserts (Dincer et al. 1974). The results are of momentous purport: first, cohesive fines bind the sand grains on the surface which forms a crust held together by fibrous and globular algae and mosses (Danin 1978, Van Den Anker et al. 1985). Second, the moisture retention

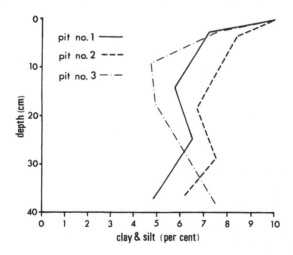

Figure 7 The amount of silt and clay (in per cent weight) measured with depth in three pits dug on semi-stabilized linear vegetated dune near the study area.

capacity of the sand is changed causing an increase in the amount of available and usable water. Third, the fine texture of the upper sand layer increases soil fertility. The above is an ecological change of the dune system that brings about a positive feedback mechanism. Fine sand containing some silt is less mobile and can support denser vegetation which, in turn, brings about additional deposition of fines thus leading finally to plant succession (Danin 1978, Kutiel et al. 1979/80, Danin & Yaalon 1982).

The contribution of fines is a phenomenon that is an exclusive feature of deserts where a great amount of dust settles down. Vegetated linear dunes in Rice Valley, California, contain fines up to a ratio of between 4.5% and 6.6%. In Australia they contain up to 5% (Breed & Breed 1979), while fines are absent in the coastal vegetated dunes of The Netherlands (Rutin 1983). Hence, vegetated sand dunes in deserts are more easily prone to stabilization, thanks to the addition of fines, than those in humid areas, where the amount of settling dust is meagre and the leaching process is very intensive. According to Yaalon and Ganor (1979), the rate of potential dust deposition in the research area has to be between 150 and 200 g m^{-2} yr^{-1}. If all the settled dust is trapped, then at least 250 years are needed to fill half a meter of sand with 4%-6% of fines. When the vegetation is destroyed and the dune becomes active, most of the fines become thereby exposed on the surface and, while the dune advances, are swiftly lost to suspension mode; the amount of fines is consequently reduced to about 1% and even less.

Direction of sand moving winds in relation to alignment of linear dunes

In many areas vegetated linear dunes are reported to align approximately parallel to the dominant winds. Secondary winds usually exert a modifying influence on the crest and account for either the symmetry or the asymmetry of the whole dune (Madigan 1946, Folk 1971, Twidale 1972, 1981, Breed & Breed 1979, Lancaster 1981). Other reports cover a near parallelism of the dunes that are concordant with the annual resultant direction (Breed & Breed 1979, Breed et al. 1979, Fryberger 1979). The Australian linear dunes display, in most cases, a small deviation from mean wind direction in their alignment (Brookfield 1970). The parallelism of linear dunes to dominant wind direction contributed to planting the idea that vegetated linear dunes are formed by helicoidal wind flow (Folk 1971, Wilson 1972). This hypothesis was developed without any reference to the vegetation cover as a determining factor in dune formation, and it was only based on secondary flows in the atmospheric boundary layer.

Figure 8 shows the weighted sand moving wind directions calculated according to Tsoar (1983), for two meteorological stations. Station 1 is located 12 km northeast of the research area and its data were collected between 1975 and 1983. Station 2 is located 10 km to the southeast and its data were collected between 1975 and 1982. Also shown in Figure 8 is the alignment of the linear vegetated dunes in the research area and the resultant vector of the wind. The two stations show a slight difference in the wind regime, but the modal direction (270°) in both cases is identical. The linear vegetated dunes deviate by 20° from the resultant vector of the wind direction, but align parallel to the modal of the tempest direction.

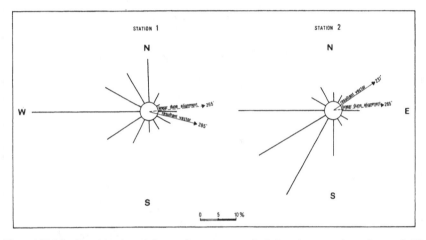

Figure 8 Weighted sand moving wind roses for two meteorological stations near the study area. Solid line arrow indicates the resultant vector direction of the wind. Dashed line arrow indicates the average vegetated linear dune alignment.

Change in the morphology of linear vegetated dunes after the destruction of vegetation

When linear dunes are covered by plants, they have a low shape with a rounded profile (Figs 3 & 6). Soon after the removal of the vegetation the shape and the profile of the dunes are subject to change. The strong wind blowing from the southwest can act freely on the dune and is responsible for the creation of the superimposed dunelets with slip faces facing northeast. Thus, the first reaction is a change from linear vegetated shape to linear braided (Fig. 5) which is asymmetrical. In some cases, a longitudinal sharp-edged crestline is formed along the latter's southern side. In those cases where linear vegetated dunes change their azimuth of alignment with an increase of 16° to 25°, which happens when they converge to form a Y-junction, they turn into seifs (Fig. 5).

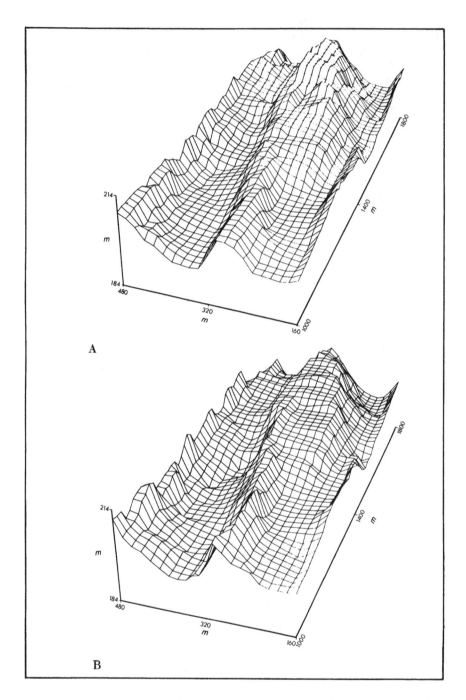

Figure 9 Three-dimensional diagram of a linear dune of the study area. A—shows the situation in 1968, B—in 1982.

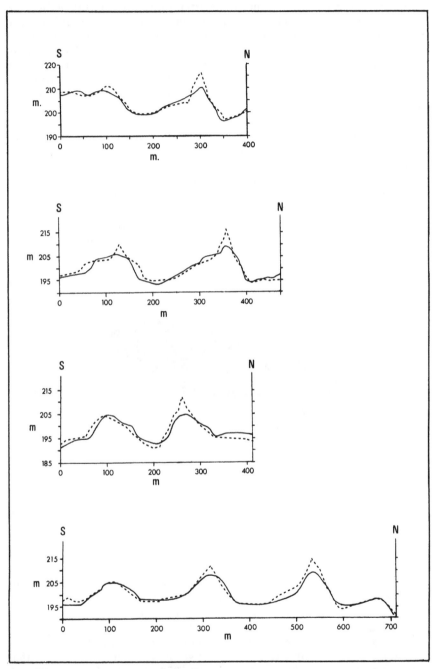

Figure 10 Four cross sections made perpendicularly to the linear dune alignment. Solid line indicates cross section taken in July 1968 and hachured line is of cross section taken along the same line in April 1982.

Figure 9 shows two 3-dimensional diagrams. A - is taken in 1968 and B - in 1982. The vegetated linear dune of 1968, in the foreground of the diagram, turned into seif with a sharp-edged crestline. A linear braided dune developed in the background. Figure 10 shows four cross sections taken perpendicularly to the linear dune alignment, in 1968 and in 1982. The permutation of linear vegetated to seif is very obvious while the change to linear braided does not stand out in the cross section, but can be discerned in aerial photographs (Fig. 5).

The maximal increase in elevation of the dune crest as a result of the permutation from linear vegetated to seif is up to 6 m. The total amount of added sand in the study area between 1968 and 1982 is 371 600 m³, equalling an average increase of surface sand of about 12 cm. The contour differences between the two maps of 1968 and 1982 were calculated and plotted in a 3-dimensional diagram (Fig. 11). It shows that the increase in elevation of the area by 1982 occurred, in most cases, along the dune crestline in those places where seif dunes developed.

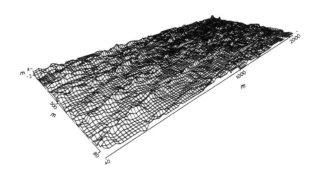

Figure 11 Three-dimensional diagram of the difference between the contours of the 1982 map and the 1968 map of the study area.

Dynamical significance of dune morphological change

The permutation of linear vegetated dune to linear braided shape can be considered as a change from a simple pattern to a complex one. The transformation into a seif dune is much more significant from a dynamical point of view. The alignment of seif dunes is not parallel to the main strong winds, as happens with linear vegetated dunes, but rather obeys two main wind directions that act upon the dune alternately (Tsoar 1978, 1983). In the research area, these two winds are westerly and southwesterly, blowing during winter time, and northerly and northwesterly in summer time

(Fig. 8). The seif dunes stretch between them in an average azimuth of 285° and 290°. The summer winds, which are of sea breeze origin, reach an average maximal value of 8 m s^{-1} while the winter winds, which are of a cyclonic origin, reach 14 m s^{-1}. The linear vegetated dunes stretch parallel to the winter winds which are the strongest in the area.

Vegetation is considered as a surface roughness element that tends to decrease the force of the wind on the sand. Hence, only strong winds are effective on vegetated dune. As emphasized before, in deserts vegetation thrives more on sand than elsewhere. A patch of sand may be covered by a dense cover of vegetation. The wind interaction with these roughness elements, which are semi-porous to non-porous, results in the formation of linear pyramidal-shaped shadow dunes in the lee side of the strongest dominant wind (Hesp 1981). More vegetation will clutch to the shadow lee dune side in a process of self-propagation, forming a linear vegetated dune in the direction of the strongest dominant wind. Cross winds will add sand to the linear dune and, in some cases, form an asymmetrical profile.

Seif dunes devoid of vegetation have a different mechanism of elongation. Winds striking the dune obliquely are separated and consequently deflected on the lee side where sand movement parallel to the crest line is reduced (Tsoar 1978, 1983). It was found, both in the field and in wind tunnel simulation work, that wind encountering the dune at an angle of incidence of about 30° ± 10° produces the maximum velocity of the deflected wind on the lee side. For that reason, seif dunes are developed in those areas (Y-junction) where vegetated dunes are aligned to angles of incidence that are proper and respond to the two main wind directions. Separation flow on the lee side is a precondition for flow deflection on the lee side as it occurs in seif dunes. Moderate slopes prevent or reduce wind separation and the ensuing deflected flow.

Models of formation of vegetation on dune sand and its destruction (desertification)

Two flow charts (Figs 12 & 13) indicate two processes which, though opposed to one another, have positive feedbacks. When sand accumulates in a desert, the better soil moisture retention it provokes attracts vegetation. As a result, the rate of sand movement and erosion decreases and less moist sand is exposed to drying up. As was mentioned before, evaporation is the main cause of moisture loss in desert soils. Vegetation and reduced sand movement also protect suspended fines settling down. This brings about the formation of an algae crust that boosts the cohesion of the grains and reduces saltation (Fig. 12). Accordingly, in desert areas with an annual average precipitation of >50 mm, dune sand tends to stablize or semi-stabilize.

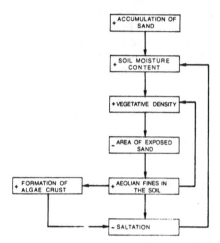

Figure 12 Flow chart of the processes and positive feedbacks of the natural stabilization of dune sand in deserts.

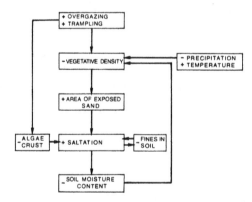

Figure 13 Flow chart of the processes and positive feedbacks following the destruction of vegetation on desert sand dunes.

Destruction of vegetation, which can be a quick process, stimulates sand movement. In linear dunes, the crest is the most exposed mobile part and hence subject to fast evaporation of the thus further contributes to the increase of sand transport and to the return of the fraction of fines to suspension in the atmosphere. Such a devolutionary process can be the consequence of climatic change or of the impact of man, as it was in the research area (Fig. 13). It may eventually result in permutation of dunes from semi-stabilized and vegetated linear shape to the much more dynamic braided and seif shapes.

Conclusions

(a) Desert areas with a perennial rainfall of more than 50 mm as annual average can, in most cases, support some vegetation on dune sand which, in turn, invigorates the deposition of dust which is plentifully available. Hence, they can be easily and naturally stabilized. The fact that many sand dunes in arid and semi-arid areas are not covered at all by vegetation is attributable to the land use history.

(b) By impeding sand movement and creating surface roughness, vegetation on desert sand dunes acts as a trap for suspended motes of silt and clay. Some of these particles are conveyed by percolating rainwater to deeper layers. The greatest amount of fine particles are concentrated on the sand surface where, with fibrous and globular algae and mosses, they create a protective crust.

(c) Evaporation is the main cause of moisture loss in desert soil. Dune sand protects its moisture as long as there is no sand erosion. For this reason, movement and exposure of sand is the major limiting factor for vegetation on sand dunes.

(d) Linear dunes form a good substratum for plants because of being elongated with little exposure of sand, unlike transverse and barchan dunes where the windward side sand is exposed constantly.

(e) Vegetated linear dunes stretch parallel to the dominant strongest wind direction.

(f) Seif dunes stretch between two dominant wind directions, forming an angle of $30° \pm 10°$ between the dune crestline and each of the two dominant wind directions.

(g) Linear braided dunes are an indication and the result of the destruction of vegetation over a linear vegetated dune.

References

Adams, R., M. Adams, A. Willens and A. Willens 1978. *Dry lands: man and plants.* London: The Architectural Press.

Bagnold, R.A. 1941. *The physics of blown sand and desert dunes.* London: Methuen.

Bagnold, R.A. 1951. Sand formation in southern Arabia. *Geographical Journal* 117, 78-86.

Bailey, C. and A. Danin 1981. Bedouin plant utilization in Sinai and

Negev. *Economic Botany* 35, 145-62.

Bailey, C. and R. Peled 1974. *The Bedouin tribes of Sinai.* (in Hebrew). Tel Aviv: Israel Defence Ministry.

Blom, R. and C. Elachi 1981. Spaceborne and airborne imaging radar observations of sand dunes. *Journal of Geophysical Research* 86, 3061-73.

Breed, C.S. and W.J. Breed 1979. Dunes and other windforms of Central Australia (and a comparison with linear dunes on the Moenkopi Plateau, Arizona). In *Apollo - Soyuz test project summary science report* Vol. 2: *Earth observations and photography*, F. El-Baz and D.M. Warner (eds.), 319-58. Washington, D.C.: National Aeronautics and Space Administration, SP-412.

Breed, C.S., S.G. Fryberger, S. Andrews, C. McCauley, F. Lennartz, D. Gebel and K. Horstman 1979. Regional studies of sand seas, using LANDSAT (ERTS) imagery. In *A study of global sand seas*, E.D. McKee (ed.), 305-97. Washington, D.C.: Geological Survey Professional Paper 1052.

Brookfield, M. 1970. Dune trend and wind regime in Central Australia. *Zeitschrift fur Geomorphologie* Supplementband 10, 121-53.

Danin, A. 1978. Plant species diversity and plant succession in a sandy area in the northern Negev. *Flora* 167, 409-22.

Danin, A. 1983. *Desert vegetation of Israel and Sinai.* Jerusalem: Cana.

Danin, A. and D.H. Yaalon 1982. Silt plus clay sedimentation and decalcification during plant succession in sands of the Mediterranean coastal plain of israel. *Israel Journal of Earth Sciences* 31, 101-9.

Danin, A., A. Weinstein and R. Karschon 1982. The synanthropic flora of new settlements on northeastern Sinai. 1. Composition and origin. *Willdenowia* 12, 57-75.

Dincer, T., A. Al-Mugrin and U. Zimmermann 1974. Study of the infiltration and recharge through the sand dune in arid zones with special reference to the stable isotopes and thermonuclear tritium. *Journal of Hydrology* 23, 79-109.

El-Baz, E. 1978. The meaning of desert color in Earth orbital photographs. *Photogrammertic Engineering and Remote Sensing* 44, 69-75.

El-Baz, F. 1984. The desert in space age. In *Deserts and arid lands*, F. El-Baz (ed.), 1-29. The Hague: Martinus Nijhoff.

Folk, R.L. 1971. Longitudinal dunes of the northwestern edge of the Simpson desert, Northern Territory, Australia, 1. Geomorphology and grain size relationships. *Sedimentology* 16, 5-54.

Fryberger, S. 1979. Dune forms and wind regime. In *A study of global sand seas*, E.D. Mckee (ed.), 137-69. Washington, D.C.: Geological Survey Professional Paper 1052.

Fryberger, S. and A.S. Goudie 1981. Arid geomorphology. *Progress in Physical Geography* 5, 420-28.

Gabriel, A. 1938. The Southern Lut and Iranian Baluchistan: Discussion. *Geographical Journal* 92, 193-210.

Goldsmith, V. 1985. Coastal dunes. In *Coastal sedimentary environments*, 2nd edn., R.A. Davis (ed.), 303-78. New York: Springer.

Hesp, P.A. 1981. The formation of shadow dunes. *Journal of Sedimentary Petrology* 51, 101-12.

Holm, D.A. 1968. Sand dunes. In *Encyclopedia of geomorphology*, R.W. Fairbridge (ed.), 973-9. New York: Reinhold.

Jennings, J.N. 1968. A revised map of the desert dunes of Australia. *Australian Geographer* 10, 408-9.

Kutiel, P., A. Danin and G. Orshan 1979/80. Vegetation of the sandy soils near Caesarea, Israel. 1. Plant communities, environment and succession. *Israel Journal of Botany* 28, 20-35.

Lancaster, N. 1981. Aspects of the morphology of linear dunes of the Namib desert. *South African Journal of Science* 77, 366-8.

Lancaster N. 1982. Linear dunes. *Progress in Physical Geography* 6, 475-504.

Lowman, P.D. 1966. The Earth from orbit. *National Geographic* 130, 645-71.

Mabbutt, J.A. 1968. Aeolian landforms in Central Australia. *Australian Geographical Studies* 6, 139-50.

Madigan, C.T. 1946. The Simpson desert expedition, 1939 scientific report: No. 6, geology - the sand formations. *Transactions of the Royal Society of South Australia* 70, 45-63.

Marx, E. 1967. *Bedouin of the Negev*. Manchester: Manchester University Press.

McKee, E.D. 1979. Introduction to a study of global sand seas. In *A study of global sand seas*, E.D. Mckee (ed.), 1-19. Washington, D.C.:Geological Survey Professional Paper 1052.

McKee, E.D. and G.C. Tibbitts 1964. Primary structures of a seif dune and associated deposits in Lybia. *Journal of Sedimentary Petrology* 34, 5-17.

McKee, E.D., C.S. Breed and L.F. Harris 1973. A study of morphology, provenance, and movement of desert sand seas in Africa, Asia, and Australia. In *Symposium on significant results obtained from the EARTS-1* Vol. 1: *Technical presentations Section A*, S.C. Freden, E.P. Mercanti and M.A. Becker (eds.), 291-303. Washington, D.C.: NASA SP-327.

Melton, F.A. 1940. A tentative classification of sand dunes and its application to the dune history in the southern High Plains. *Journal of Geology* 48, 113-74.

Muehlberger, W.R. and V.R. Wilmarth 1977. The shuttle era: A challenge to the Earth scientist. *American Scientist* 65, 152-8.

Noy-Meir, I. 1973. Desert ecosystems: environment and producers. *Annual Review of Ecology and Systematics* 4, 25-51.

Noy-Meir, I. and N.G. Siligman 1979. Management of semi-arid ecosystems in Israel. In *Management of semi-arid ecosystem*, B.H. Walker (ed.), 113-60. Amsterdam: Elsevier.

Olson, J.S. 1958. Lake Michigan dune development 2. Plants as agents and tools in geomorphology. *Journal of Geology* 66, 345-51.

Otterman, J. 1974. Baring high-albedo soils by overgrazing: a hypothesized desertification mechanism. *Science* 186, 531-3.

Otterman, J. 1981. Satellite and field studies of man's impact on the surface in arid regions. *Tellus* 33, 68-77.

Otterman, J., Y. Waisel and E. Rosenberg 1975. Western Negev and Sinai ecosystems: comparative study of vegetation, albedo and temperatures. *Agro-Ecosystems* 2, 47-59.

Price, W.A. 1950. Saharan sand dunes and the origin of the longitudinal dune, a review. *Geographical Review* 40, 462-5.

Pye, K. 1982a. Morphological development of coastal dunes in a humid tropical environment, Cape Bedford and Cape Flattery, North Queensland. *Geografiska Annaler* 64A, 213-27.

Pye, K. 1982b. Thermo-luminescence dating of coastal dunes. *Nature* 299, 376.

Rutin, Y. 1983. *Erosional processes on a coastal sand dune, De Blink, Noordwijkerhout, The Netherlands.* Unpublished Ph.D. Thesis. University of Amsterdam.

Schneider, S.H. 1976. *The genesis strategy.* New York: Plenum Press.

Smith, H.T.U. 1963. *Eolian geomorphology, wind direction, and climatic change in North Africa.* Bedford: Air Force Cambridge Research Laboratories AFCRL-63-443.

Streim, H.L. 1954. The seifs on the Israel-Sinai border and the correlation of their alignment. *Bulletin of the Research-Council of Israel* 4, 195-8.

Tsoar, H. 1976. Characterization of sand dune environment by their grain-size mineralogy and surface texture. In *Geography in Israel,* D.H.K. Amiran and Y. Ben-Arieh (eds.), 327-43. Jerusalem: Israel National committee, I.G.U.

Tsoar, H. 1978. *The dynamics of longitudinal dunes.* London: Final Technical Report, European Research Office U.S. Army.

Tsoar, H. 1983. Dynamic processes acting on a longitudinal (seif) sand dune. *Sedimentology* 30, 567-78.

Twidale, C.R. 1972. Evolution of sand dunes in the Simpson Desert, Central Australia. *Transactions of the Institute of British Geographers* 56, 77-109.

Twidale, C.R. 1981. Age and origin of longitudinal dunes in the Simpson and other sand ridge deserts. *Die Erde* 112, 231-41.

Van Den Ancker, J.A.M., P.D. Jungerius and L.R. Mur 1985. The role of algae in the stabilization of coastal dune blowouts. *Earth Surface Processes and Landforms* 10, 189-92.

Warren, A. 1972. Observation on dunes and bimodal sands in the Tenere Desert. *Sedimentology* 19, 37-44.

Warren, A. and C.M. Harrison 1984. People and the ecosystem: biogeography as a study of ecology and culture. *Geoforum* 15, 365-81.

Wilson, I.G. 1972. Aeolian bedforms - their development and origins. *Sedimentology* 19, 173-210.

Wilson, I.G. 1973. Ergs. *Sedimentary Geology* 10, 77-106.

Yaalon, D.H. and E. Ganor 1979. East Mediterranean trajectories of dust-carrying storms from the Sahara and Sinai. In *Saharan dust*, C. Morales (ed.), 187-93. Chichester: Wiley.

Zohar, Y., G. Schiller and R. Karschon 1971. Determination of the direction of prevailing wind from the orientation of crescent-shaped Bedouin shelters on northeastern Sinai. *Agricultural Meteorology* 8, 319-23.

6

Geomorphological significance of wind flow patterns over a Namib linear dune

Ian Livingstone

Abstract

This paper reports that part of a wider study of the dynamics of a linear dune in the Namib sand sea which is concerned with the wind flow patterns over the dune. Results of wind speed measurements using anemometers and flow visualization using smoke flares demonstrate that the dune's intrusion into the boundary-layer creates a pattern of wind flow change over the dune cross profile. By assuming that increasing wind speed causes erosion of the dune surface and decreasing wind speed causes sand deposition, a simple, qualitative model of linear dune dynamics is postulated. Without invoking secondary wind flows as previous theories have done, maintenance of linear dune form is seen as a response to the pattern of wind speed change alone.

Introduction

Despite the dominance of dunes with a linear form in the world's sand seas (Fryberger & Goudie 1981), they remain "perhaps the most controversial type of dune" (McKee 1979), the subject of a number of highly speculative interpretations and relatively few careful field studies. In recent years, however, single dune studies, notably those of Knott (1979) and Tsoar (1978), have greatly extended our knowledge of processes on aeolian dunes. Furthermore, at the same time as geomorphologists studying the development of desert dunes have become interested in wind flow over dunes, meteorologists have become concerned with the parallel problem of the effect which the intrusion of low hills into the boundary-layer has on wind fow, and Lancaster (1985) and Tsoar

(1985) have recently made use of formulae provided by meteorologists to model wind flow over dunes. There has consequently been some progress towards empirical modelling of linear dune dynamics.

Against this background an investigation has been undertaken of the dynamics of a single, complex linear dune in the Namib Desert in southern Africa (Livingstone 1985). This paper reports that part of the study concerned with wind flow patterns over the dune, and seeks to propose a simple qualitative model of linear dune dynamics.

Past theories

In the past, a number of theories have been proposed to explain linear dune origin, evolution and maintenance. There has been a difference of opinion between those believing in the role of prevailing winds blowing parallel to the dunes and those invoking necessarily bi-directional wind regimes. Much has been made of the importance of antecedent dune forms such as barchans, zibars and parabolic dunes. There has also been some discussion, reviewed in Lancaster (1982), between those invoking an erosional 'wind rift' origin and those proposing a depositional origin for linear dunes.

Two theories in particular, however, have attempted to explain the actual mechanism by which linear dunes form and advance, and it is of value to consider these theories further. The first of these, and the one which appears most often in textbooks, is the helical roll vortex theory of Bagnold (1953). In support of his theory, Bagnold cites laboratory experiments by Brunt (1937) in which polygonal cells were created by heating a metal plate. When a glass plate was then moved above the heated metal plate a shear in the boundary-layer was caused and the polygonal cells were moved laterally producing spiral roll vortices. The two essentials, therefore, for the creation of roll vortices are thermal convection and a strong geostrophic wind, conditions often found in desert areas, and Bagnold tentatively suggests this as a possible mechanism of linear dune genesis.

While this theory is attractively simple, a number of objections have been raised. There is no doubt that roll vortices do exist, but they have yet to be described from linear dune fields, and those which have been reported from other areas appear to display transverse velocities well below that required to move sand. Additionally, while Kelly (1984) reports roll wavelengths between 1.5 and 13.7 kilometers, linear dune spacing varies between 0.15 and around 3.5 kilometers (Breed & Grow 1979). Given that roll vortices are paired, dune wavelengths of between 3.0 and 27.4 kilometers could be inferred, and clearly there is little overlap between the two size distributions. The theory also requires that winds blow parallel to the dune trend. This occurs rarely in the

Namib Desert, and Besler (1980) overcomes the problem by viewing the dunes as relics of a much windier episode of winds parallel to the dune alignment at the time of the Last Glacial maximum. Yet the fact that roll vortices have still to be demonstrated creating linear dunes in any dune field, when coupled with the circumstantial evidence outlined above, suggests that it is necessary to entertain the possibility of some other mechanism of formation for linear dunes.

More recently, though, there has been a challenge to this first theory from Tsoar (1978, 1983) whose own theory describes the importance of the separation and deflection of flow in lee side eddies on linear dunes. Following the study of a simple linear dune in the Negev Desert, Israel, Tsoar proposed that,

"The basis of the dynamics of the longitudinal dune is the phenomenon that the path of the wind flow when crossing the crest at any angle whatsoever is deflected on the lee flank in the direction parallel to the crest line". (Tsoar 1978, p. 133).

In other words, Tsoar's proposal is that once formed, the intrusion of the linear dune into the boundary-layer creates eddies condusive to its own self-perpetuation, such that sand is prevented from leaving the lee side of the dune. Two conditions must be fulfilled for this mechanism to operate: there must be a bi-directional wind regime, and the dune must have a sharp crestline so that flow separation occurs in the immediate lee of the crest. Clearly, the theories of both Bagnold and Tsoar invoke some form of secondary flow, but the attraction of Tsoar's theory is that the secondary flow is caused by the dune itself rather than by atmospheric conditions.

The study dune

The study site lies at the northern edge of the Namib sand sea in Namibia, approximately eight kilometers southeast of the Namib Desert Research Station at Gobabeb (23° 34'S 15° 03'E). The dune here is aligned roughly north-south, and is approximately 350 m wide and 50 m high. The study dune displays the asymmetry typical of dunes in this part of the Namib sand sea, with a relatively uniform west flank, but a series of secondary ridges and barchanoid features on the east flank.

The central Namib Desert is subjected to a seasonal wind regime so that, broadly speaking, the dunes are attacked by low to moderate winds from the southwest and northwest in summer, and by high magnitude, low frequency easterly winds in winter (Lancaster et al. 1984). Under the influence of this regime the crest of the dune moves laterally back and forth some fourteen meters each year (Livingstone 1985), while northward extension of the dunes into the Kuiseb River valley has been measured by Ward at between 0 and 1.85 m yr^{-1} (Ward 1984).

Wind flow measurements

In order to monitor the pattern of wind flow near the dune surface two techniques were employed. Smoke flares were used to enable visualization of the flow patterns, and rotating-cup anemometers of the 'wind run' type were used to measure wind speed near the dune surface. All anemometers were mounted on posts one meter above the ground surface, and readings were usually made every 30 minutes, although during periods of relative calm, readings were made every hour. Because there were never more than five anemometers available at any one time, different dune or interdune elements were monitored at different times, but on each occasion when anemometers were located on the dune, an anemometer was placed at the crest of the dune, thereby enabling the formulation of a relationship between the wind speeds at various points over the dune cross-profile.

A total of 69 readings of winds originating from the west side of the dune and 23 readings in easterly winds were recorded at the study dune. Forty-five readings were recorded for wind speeds on the interdune corridors adjoining the study dune in both easterly and southwesterly winds. From these data it has been possible to

Table 1 Wind speeds on the study dune in south-westerly winds:
correlation and regression with the speed at the crest.

y	n	r	t	p	m	c
Widc	47	0.8743	12.1	0.001	0.72	−0.42
B	19	0.9981	66.8	0.001	0.65	−0.16
E	23	0.9969	58.2	0.001	0.80	−0.24
H	23	0.9940	41.6	0.001	0.68	+0.44
M/N	21	0.8237	6.3	0.001	0.70	−2.38
O	21	0 8986	8.9	0.001	0.95	−3.34
P	24	0.7648	5.6	0.001	0.58	−0.44
Q	21	0.9620	15.4	0.001	0.93	−2.55
S	21	0.9459	12.7	0.001	0.84	−2.22
Eidc	24	0.6077	3.6	0.001	0.64	−0.59

y   point on the dune (Widc = West interdune corridor, Eidc = East interdune corridor)
n   sample size
r   Product Moment Correlation Coefficient
t   Student's t value
p   significance level expressed as probability that the distribution is random
m   gradient of regression equation
c   intercept of regression equation

calculate the 'Product Moment Correlation Coefficient' and 'Least Squares' regression equations indicating the relationship between wind speeds at the dune crest and the speed on other parts of the dune, and thereby providing a simple linear regression model of wind speed change at the surface of a linear sand dune (Tables 1 & 2). The letters denoting location of the anemometers use the notation of a simultaneous erosion pin project.

Table 2 Wind speeds on the study dune in easterly winds:
correlation and regression with the speed at the crest.

y	n	r	t	p	m	c
Widc	19	0.8718	7.3	0 001	0.50	−1.29
A	5	0.1814	0.3		not statistically significant	
D	6	0.7594	2.3		not statistically significant	
F	5	0.7637	2.0		not statistically significant	
I/J	6	0.1558	0.3		not statistically significant	
N	5	−0.0690	−0.1		not statistically significant	
O	3	0.6032	0.8		not statistically significant	
Q	5	0.5828	1.2		not statistically significant	
R	14	0.8630	5.9	0.001	0.60	+0.55
S	7	−0.2470	−0.6		not statistically significant	
Y	9	0.9714	10.8	0.001	0.47	−0.57
Eidc	14	0.9847	19.6	0.001	0.38	+0.98

y   point on the dune (Widc = West interdune corridor, Eidc = East interdune corridor)
n   sample size
r   Product Moment Correlation Coefficient
t   Student's t value
p   significance level expressed as probability that the distribution is random
m   gradient of regression equation
c   intercept of regression equation

Considering summer winds originating on the west side first, there are very strong, positive, linear relationships between wind speeds at one point on the dune and speed at another (Table 1), and from these empirically formulated linear relationships it has been possible to calculate the wind speed over the study dune's cross-profile for given initial winds speeds on the interdune corridor (Fig. 1). From these calculations a distinctive pattern of changing wind speed on the linear dune emerges. This pattern shows a very obvious, progressive increase of wind speed from the interdune corridor to the dune crest due to a compression of streamlines, and a similar decrease from the crest to the lee side interdune corridor as the streamlines diverge. However, there is some finer detail on this overall trend for there is a very rapid drop in wind speed in the immediate lee of the crest where the slip face is formed. This is the result of the separation of the flow from the ground surface described by Tsoar (1978), and is discussed in the following section on 'Flow Visualization'. The wind then

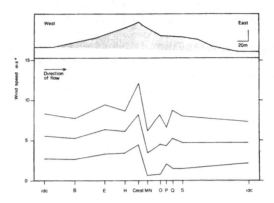

Figure 1 Wind speeds calculated from the regression equations for flow over the study dune in south westerly winds.

recovers speed away from the base of the slip face, represented by readings from point M/N, to the secondary dune crest, represented by readings from Q and S. Between M/N and Q, there is again a slightly anomalous pattern introduced by the series of troughs and barchanoid forms in this area of the east slope of the study dune.

Unfortunately, the strong relationships between speeds at different points on the dune found for westerlies are not repeated in the results for easterly winds (Table 2). This may well be because the easterly winds, as well as being stronger, are also 'gustier' than the southwesterly winds, and there is therefore likely to be far greater variability in their readings. In addition, there are fewer readings from easterly winds, so that the effect of variability on the calculations is increased. Furthermore, poor correlation is associated with zones where there are turbulent eddies on the lee side, west slope, and in the lee of the secondary dune ridge.

Despite the lack of significant statistical relationships between points on the dune surface in easterly winds, it is possible to present five individual transects from winter, 1981, and to describe qualitatively the pattern they show (Fig. 2). Again, the speed at the crest is very clearly greater than the speed in either interdune corridor. There are, however, very sharp decreases in speed in the lee of the main dune crest and the lee of the east slope secondary dune. At the crest, the difference is between 17.5 m s^{-1} at the crest and 1.7 m s^{-1} ten meters downwind on one transect, and between 22.8 m s^{-1} and 3.9 m s^{-1} on another. This represents a very great decrease in the competence of the wind to transport sand. The greater variations of the speeds recorded on the lee slope compared with the range of initial speeds at the windward base are a manifestation of the turbulence and the existence of

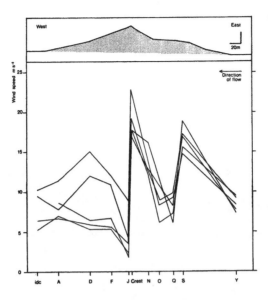

Figure 2 Five sets of readings of wind speeds over the study dune in easterly winds.

eddying on the lee slope.   The data from the interdune corridors for winds from a variety of directions indicate that wind speed across the corridors at one meter above the ground surface is more or less constant.

Flow visualization

To support the data collected from the anemometers, smoke flares were used to visualize the flow pattern over the dune.   The release of smoke flares of the kind used as ground targets for parachutists provides a qualitative impression of wind flow patterns which can be recorded photographically.   The technique has proved highly successful for the visualization of eddies, and a summary of the results of the flow visualization investigation are presented in diagramatic form in Figure 3.

While Tsoar (1978) found that the lee side eddy covered the entirety of the lee slope of his study dune, it is clear from the smoke flares that this is not the case on the study dune of the present project.   Figure 3 represents the situation in a southwesterly wind and shows that, while a lee side eddy does exist, the eddy covers only the upper part of the east slope.   On the lower slope the flow appears to have re-attached to the surface of the dune so that flow on the lower lee slope is also southwesterly.   The clear consequence of this observation is that

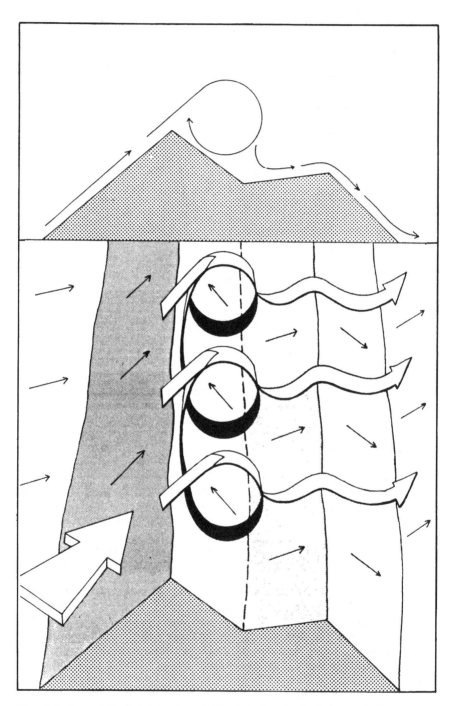

Figure 3 Summary of the wind flow pattern over the study dune visualized using smoke flares.

on a 'complex' dune the size of the present study dune, the lee
side eddy does not prevent sand from leaving the dune and
crossing the interdune corridor. In other words, Tsoar's
postulated mechanism for linear dune maintenance of a lee side
eddy in which net movement of sand is longitudinal does not seem
to be universally applicable.

Towards a model of linear dune dynamics

It has proved possible to demonstrate, then, that there are
significant changes in the speed of the near-surface wind over a
linear dune cross-profile, and to provide an empirically derived
model of patterns of wind speed change for westerly winds on the
study dune. Naturally, it would be of considerable advantage to
both pure and applied studies of dune dynamics to be able to model
the relationship between the wind flow patterns outlined above and
the dynamics of sand transport at the dune surface. Wind speed
alone, however, is a measure only of the capacity of the wind to
carry sand. Landform change - erosion and deposition - is effected
by variations of that capacity related to changes of wind speed in
both time and space, and a simple intuitive model can be proposed
whereby positive acceleration gives erosion, negative acceleration
gives depositon, and zero acceleraton gives no net change in the
dune form (Cooke & Warren 1973, Holm 1960).

The changes in the wind's speed and direction on an individual
dune are caused by effects at two scales superimposed one on the
other. In both cases, it is the role of converging or diverging
airflows which have been emphasized. At the macro-scale of an erg,
the regional wind changes speed and direction in response to the
regional pressure gradient. However, in a single dune study such
as this covering a relatively small area, a constant regional wind
speed can be assumed. At the second scale, which is more
important to the behaviour of individual dunes, there is a pattern
of wind speed change across the dune profile. Wind speed varies
as a response to the convergence and divergence of streamlines
caused by the intrusion of the dune into the wind flow. It should
also be noted, however, that there exists an important inhibition to
sand movement on the windward slopes where the increased
gradient of the dune flank will increase the effective force due to
gravity countering upslope movement, and a greater wind speed
would be required to entrain the sand (Howard 1977).

Nonetheless, setting aside considerations of gradient, and
assuming a regional wind of constant velocity, this qualitative
model in conjunction with the regression equations for wind speed
over the study dune presented in Table 1 outlines a
two-dimensional description of surface change across the linear
dune profile. Figures 4 and 5 show a schematic representation of
the topography of the study dune, the inferred patterns of wind

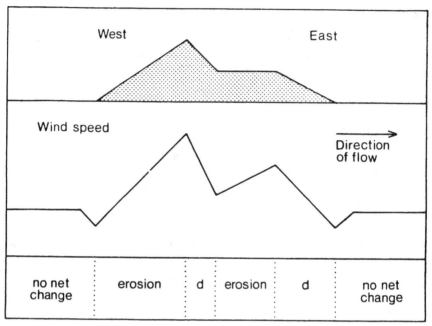

Figure 4 Relationship between wind speed and surface change in winds originating from the west side of the study dune.

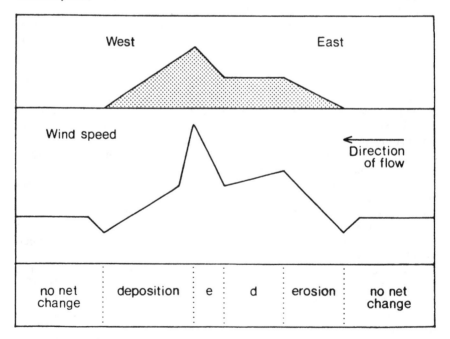

Figure 5 Relationship between wind speed and surface change in winds originating from the east side of the study dune.

speed, and surface changes in south-westerly winds (Fig. 4) and easterly winds (Fig. 5). Without recourse to any indication of wind flow direction, the model of wind speed over the dune cross-profile can be extended to predict the salient dynamic features of the dune. It shows erosion of the windward flank, deposition on the slip face, erosion of the zone to the lee of the slip face, deposition on the lee flank, and no net change on the interdune corridor. When these deduced patterns of surface change are compared with the actual patterns recorded from erosion pin data, a very good correlation of the overall trends is apparent (Livingstone 1985). From these patterns it is possible to propose a model of linear dune dynamics whereby the dune's intrusion into the atmospheric boundary-layer leads to its own self-perpetuation.

It appears that wind velocities on the corridors on either side of the dune are more or less the same. Assuming an equal supply of sand, the volume of sand carried across the two adjacent interdune corridors is therefore the same, and the volume of sand arriving at the upwind base of the dune is matched by the volume of sand leaving at the downwind base. The total volume of sand in the dune therefore remains (roughly) constant. Sand may be eroded from the windward flank and deposited on the lee flank thereby leading to lateral movement, but there is no net loss or gain of sand, and all changes of form occur as a result of sand budget variations on the dune itself. This deduction of no net erosion or deposition on the interdune corridors is important, for if there is no net change, interdune corridors once formed are maintained. As a result there is no necessity in this model to invoke any aerodynamic control of the interdune corridors as, for instance, Tsoar (1978) has done. While Tsoar's model seeks to prevent sand leaving the dune, the present model allows sand to leave the dunes to cross the interdune corridors, but because of the constant wind velocity, there is no net change of the interdune surface.

The wind speed measurements from the interdunes also have some significance for the 'windbreak' theories of linear dune spacing proposed by Tsoar (1978) and Twidale (1972). Twidale's argument is that on the lee side, separated flow covers the entire interdune corridor, and that a neighbouring dune forms at the point of re-attachment of flow. Neither the present study nor Tsoar's provide support for lee side eddies of this size (although it is possible that some mechanism of this sort might be responsible for the 'compound' linear dunes described by Lancaster (1982)). Tsoar's theory proposes that the spacing between dunes is a function of the distance required for the wind to recover its initial speed, and that dunes form at the point when this initial speed is regained. It is not clear how this mechanism might operate, and the data from the present study of wind speeds on the interdune corridors provide no evidence to support a theory proposing gradual recovery of speed across the interdune corridors. Rather, it seems that wind speed recovers near the downwind base of the dune and is constant across the interdune corridor. Based on these results, a 'windbreak' hypothesis of linear dune spacing cannot be supported.

A model in which it is the intrusion of the dune into the boundary-layer which leads to its own self-perpetuation and in which the interdune spaces are geomorphologically by-passed is not exclusively applicable to linear dunes. Indeed, were the wind regime unidirectional, erosion on the upwind slope and deposition on the downwind slope would lead to a downwind migration of the dune with the crest normal to the incident wind direction: the dune would be developing as a transverse dune. But as a result of the bi-directional wind regime (or more accurately a non-uni-directional wind regime) zones of erosion in winter become zones of deposition in summer and vice versa, and the dune does not progress like a transverse dune, but oscillates about a central position.

Both the present study and the study of Tsoar (1978) have shown that as the wind crosses the crest there is a deflection of the flow direction along the dune causing some longitudinal sand movement, and this is the dynamic process which Tsoar invokes to explain the maintenance of linear dune form and the absence of lateral movement. According to Tsoar, the separation of flow and the lee side eddy are the key to linear dune dynamics. Yet on a dune of the size of the present study dune, it has been demonstrated that this may not be entirely applicable. There is clear evidence, both from smoke flares and from the present author's observations of surface sand movement, that the lee side eddy does not cover the entire east slope in the moderate winds originating from the west side. Because lee side eddies do not cover the entire lee flank of a dune of the size of the study dune, a model which emphasizes the importance of wind speed in dune development has advantages over a model, such as Tsoar's (1978), which emphasizes the role of wind direction, particularly of the diversion of flow in lee side eddies. A model based on wind speed is applicable to dunes of any size.

Figure 6a illustrates a simplified representation of the situation in the northern Namib sand sea. The deduced, seasonal movement of the dune tip is similar to the pattern described by Besler (1975, 1980) for the movement of a simple, linear dune close to the site of the present study, and to the pattern of advance of complex dune tips monitored by Ward (1984) in the Kuiseb River. In the hypothetical situation of a regime in which the angle between the modes is small (Fig. 6b) advance of the tip of the dune would be more rapid. This model can be extended. Were the volume of sand supplied to the two dunes in Figure 6 the same, because of the greater rate of advance in case b, a lesser volume of sand in the cross-sectional area could be deduced, and a smaller dune would be formed. It is proposed, therefore, that linear dune cross-sectional area (height and width) is a dynamic equilibrium response to the interaction of wind regime (direction and strength) and the volume of sand available. Thus in a regime in which sand supply is plentiful, but the modes of the wind regime are widely divided, the tendency will be for slow extension, and larger dunes will be built. In a model of this nature, once an equilibrium form is reached, time is irrelevant to the dune dimensions of height and width, but is relevant to the distance that the dune has extended from the source of sand.

This model of linear dune dynamics based on the pattern of wind speed change over the dune fails to elucidate two major problems. The first, common to all dune studies, concerns the mode of dune initiation. In his discussion of barchans Knott (reported in Warren & Knott 1983) has highlighted the role of converging air flow in thermally generated 'ground jets'. Clearly, further investigation is required to examine the conditions, both meteorological and topographic, by which the wind's capacity to carry sand drops and sand is deposited. The second major problem, which is particular to linear dune studies, is that of dune spacing (or wavelength). While the spacing in some linear dune fields displays a wide range of values and may therefore be some response to wind regime and sand supply, it is equally true that there are areas of linear dunes where spacing is outstandingly regular, and it is difficult to avoid the inference that this is a response to some secondary flow or wave pattern in the wind. If this secondary flow control of dune spacing exists, it remains to be discovered.

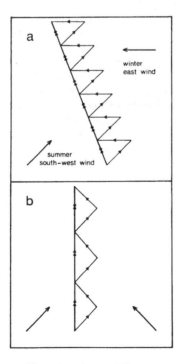

Figure 6 Inferred pattern of advance of linear dunes in two bi-directional wind regimes.

Conclusion

This paper has outlined the results of an investigation of wind flow patterns over a complex linear dune, and proposed a model of dune form maintenance based on the pattern of wind speed change over the dune cross-profile.   Compression of streamlines on a rising, upwind slope leads to acceleration of the wind and consequently to erosion.   Divergence of streamlines on the downwind slope leads to a decelerating wind and sand deposition.   By this mechanism, dunes migrate downwind, yet the amount of sand arriving from the interdune corridor at the upwind base of the dune is matched by the sand leaving the downwind base, so that the net volume of sand in the dune is maintained.   Linear dunes are viewed as essentially two or more transverse dune systems superimposed on the same sand body as a response to a non-uni-directional wind regime.   Downwind advance in one direction is countered by movement in another and net dune advance is along some resolution of the vectors of wind speed and direction.   In this model, linear dune form is seen as a response to the volume of sand   supplied and the wind regime (Wasson & Hyde 1983).   Unlike Bagnold (1953) and Tsoar (1978), no secondary wind flow is invoked and the only necessary condition for linear dune formation is seen as a non-uni-directional wind regime.
    Clearly, this model is very far from being comprehensive.   It fails, for instance, to take account of the linear dunes which appear to be related to regimes in which wind flow is predominantly from one direction parallel to the dune's alignment (Lancaster 1982),and it may well be that the linear dune represents an equifinal form.   The problem of whether dunes are truly contemporary forms or are largely a response to antecedent conditions also remains to be tackled.
    This study, along with those of Knott (1979) and Tsoar (1978), has shown that much can be discovered from process studies of individual dunes.   There still exists, however, considerable scope for more information to be gleaned from careful field investigations of the relationship between wind regime and linear dune form in a variety of dune fields.   When coupled with the formulae developed by meteorologists for wind flow over topographic barriers, this information will enable us to move closer to an understanding of linear dune dynamics.

Acknowledgements

The funding for the project came from a Natural Environment Research Council (U.K.) Research Studentship held at the School of Geography, University of Oxford. Permission to work in the Namib/Naukluft National Park was granted by the Department of Nature Conservation, Namibia, and facilities at the Namib Desert Research Station were made available through the Desert Ecological Research Unit. I am grateful to Andrew Goudie and Bob Moore for their comments on a draft of this paper.

References

Bagnold, R.A. 1953. The surface movement of blown sand in relation to meteorology. *Research Council of Israel Special Publication* 2, 89-96.

Besler, H. 1975. Messungen zur Mobilitat von Dunensanden am Nordrand der Dunen - Namib (Sudwestafrika). *Mitteilungen der geographischen Gesellschaft, Wurzburg* 43, 135-47.

Besler, H. 1980. Die Dunen - Namib: Entstehung und Dynamik eines Ergs. *Stuttgarter Geographische Studien* 96.

Breed, C.S. and T. Grow 1979. Morphology and distribution of dunes in sand seas observed by remote sensing. *United States Geological Survey Professional Paper* 1052, 253-302.

Brunt, D. 1937. Natural and artificial clouds. *Quarterly Journal of the Royal Meteorological Society* 63, 277-88.

Cooke, R.U. and A. Warren 1973. *Geomorphology in deserts*. London: Batsford.

Fryberger, S. and A.S. Goudie 1981. Arid geomorphology. *Progress in Physical Geography* 5, 420-8.

Holm, D.A. 1960. Desert geomorphology in the Arabian peninsula. *Science* 132, 1369-79.

Howard, A.D. 1977. Effect of slope on the threshold of motion and its application to orientation of wind ripples. *Bulletin of the Geological Society of America* 88, 853-56.

Kelly, R.D. 1984. Horizontal roll and boundary-layer interrelationships observed over Lake Michigan. *Journal of Atmospheric Science* 41, 1816-26.

Knott, P. 1979. *The structure and pattern of dune-forming winds*. Unpublished Ph.D. Thesis, University of London.

Lancaster, J., N. Lancaster and M.K. Seeley 1984. Climate of the central Namib Desert. *Madoqua* 14, 5-61.

Lancaster, N. 1982. Linear dunes. *Progress in Physical Geography* 6, 475-504.

Lancaster, N. 1985. Variations in wind velocity and sand transport on the windward flanks of desert sand dunes. *Sedimentology* 32, 581-93.

Livingstone, I. 1985. *The dynamics of sand transport on a Namib linear dune.* Unpublished D.Phil. Thesis, University of Oxford.

McKee, E.D., (ed.) 1979. *A study of global sand seas.* United States Geological Survey Professional Paper 1052.

Tsoar, H. 1978. *The dynamics of longitudinal dunes.* Final Technical Report, European Research Office, United States Army, London, DA-ERO 76-G-072.

Tsoar, H. 1983. Dynamic processes acting on a longitudinal (seif) dune. *Sedimentology* 30, 567-78.

Tsoar, H. 1985. Profile analysis of sand dunes and their steady state signification. *Geografiska Annaler* 67A, 47-59.

Twidale, C.R. 1972. Evolution of sand dunes in the Simpson Desert, central Australia. *Transactions, Institute of British Geographers* 56, 77-109.

Ward, J.D. 1984. *Aspects of the Cenozoic geology in the Kuiseb valley, central Namib Desert.* Unpublished Ph.D. Thesis, University of Natal.

Warren, A. and P. Knott 1983. Desert dunes: a short review of needs in desert dune research and a recent study of micro-meteorological dune-initiation mechanisms, In *Eolian sediments and process*, M.E. Brookfield and T.S. Ahlbrandt (eds.), 343-52. Amsterdam: Elsevier.

Wasson, R.J. and R. Hyde 1983. Factors determining desert dune type. *Nature* 304, 337-9.

7

The Toshka-Canal dune: analysis of development and dynamics

Helga Besler

Abstract

The Toshka Canal was built as a spillway from Lake Nasser (Egypt) leading into the Toshka Depression. In order for construction to be carried out, it was necessary to cut and remove the central section of a single dune ridge which crossed the proposed canal route. To understand the dynamics of this longitudinal dune and the future behaviour of the two remaining portions, an investigation was carried out during 1982, including analysis of the interior (textural and structural, dating) and of surface changes (tachymetric and granulometric survey, surface dyeing, measurements of sand movement, photographic records and air photo interpretation). These combined techniques revealed a complex dune history and the changing behaviour of the present dune.

The first precursors were small barchan dunes which degenerated into vegetated sand shields during a moister period around 1800 years B.P. Later, these features trapped blown sands and were finally overrun by the present lee dune forming behind a prominent section of a small escarpment to the north. Today, the dune consists of a stable base and stable flanks which act as transportation surfaces for blown sand resulting in the extension of the dune and the active shifting crest. The northern dune section is breeding barchans at its truncated southern position; this has been triggered by human activities. Where the winds meet the man-made southern portion of the northern dune section, a corresponding barchan structure has developed on top.

Introduction

The Toshka Canal was built as a spillway for Lake Nasser. It was completed in 1982 and connects the southern part of the lake with the Toshka Basin south of the Kharga Depression (Fig. 1). The canal was constructed along the Khor Toshka, a former (late middle Pleistocene) tributary of the Nile river north of Abu Simbel, which is now a shallow dry valley containing some 30 m of unconsolidated sediments (Haynes 1980).

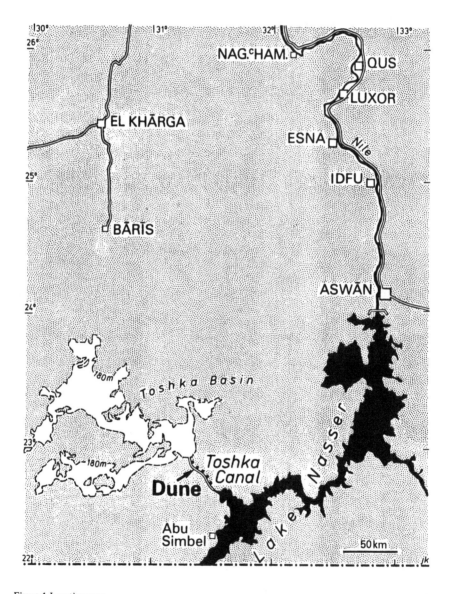

Figure 1 Location map.

A single longitudinal dune ridge up to 20 m in height crossed the Khor at almost right angles. Construction resulted in the removal of the dune's central portion such that the Toshka-Canal dune now consists of two parts: a northern dune ridge ending north of the north bank, and a southern dune ridge starting on the south bank. The dynamics of these dune parts as a possible hazard for the canal were studied.

Dune topography

Figure 2 View across the Toshka Canal with its shelter belt of young trees, looking SSE from the northern part of the dissected Toshka-Canal dune. Note the shield dune in front and the longitudinal dune ridge running SSW in the distance, reaching the south bank of the Khor Toshka at the horizon. (Photo: H. Besler, 10/3/82.)

The Toshka-Canal dune (Fig. 2) begins as a lee dune behind a prominent scarp; approximately 20 m in height located on the northern bank of the Khor. The dune ridge is rather symmetrical with both flanks inclined at about 20°. This lee dune, is oriented towards the SSE. This results from the fact that the scarp is alinged to the west at the start of the dune and therefore produces a strong westerly wind flow. Once the dune approaches the open area of the Khor, it changes direction and runs SSW according to the prevailing NNE winds and the height declines from 18 to 13 m; it is in this section that the dune was cut for construction of the canal. The main dune body is situated south of the canal and rises from 14 m to 20 m in elevation. The crest line swings laterally as well as vertically giving it a cock's-comb-like appearance. This double sinuosity coincides with undulations on the flanks. The gradient of the eastern slope is less steep and averages 10° in the lower and 15° in the higher reaches. The dune ends on the south bank of the Khor Toshka in the form of a small sand ramp descending into a southern depression.

Sand for the dune is supplied from the low, dissected plateau north of the Khor where sand is transported by winds which are locally channeled by scarp topography, and accumulates where the divided lines of air flow meet again. The dune body was obviously growing downwind until it reached the southern bank of the Khor where sand was piling up, but the top of the rocky bank is kept clear of sand because of a sudden increase in the rate of transport due to stronger winds at this elevated part. Evidence is provided by the presence of deep and fresh wind fluting in the rock.

Dune development

In order to acquire knowledge about the formation of the dune ridge and to assess the possible future growth of its remaining sections, the interior of the dune was studied. The most favourable site for this work was found to be the northern end of the southern dune section; it is here that two cuts were made successively at right angles to each other (200° and 290°) by a bulldozer.

Structural analysis

Since the sand dried quickly and the walls tended to collapse, the analysis of bedding planes was restricted to the western half of the dune body. Here, the most interesting section of the whole sequence proved to be the lower profile. Since the upper profiles displayed very homogeneous bedding structures, investigations focussed on the first few meters above ground.

Figure 3 shows the dip of sand strata measured within two profiles 3 to 4 m apart in the main cut approximately perpendicular to the axis of the dune (290°), and within one profile of the side cut (200°). All bedding planes dip in the same direction with little or no cross bedding. Fortunately, a conspicuous layer of sand mixed with organic matter is present in all three profiles which allows for the correlation of strata. The true dip angles and directions were calculated from the combined profiles, and are listed to the right of Figure 3.

The dune rests on a pale reddish sandy soil also found in the vicinity. It contains numerous shells of land snails and is covered by a thin desert pavement which also extends beneath the dune. Obviously, the surrounding land surface has not changed much since dune formation. The first lower sand layers dipping towards SSW

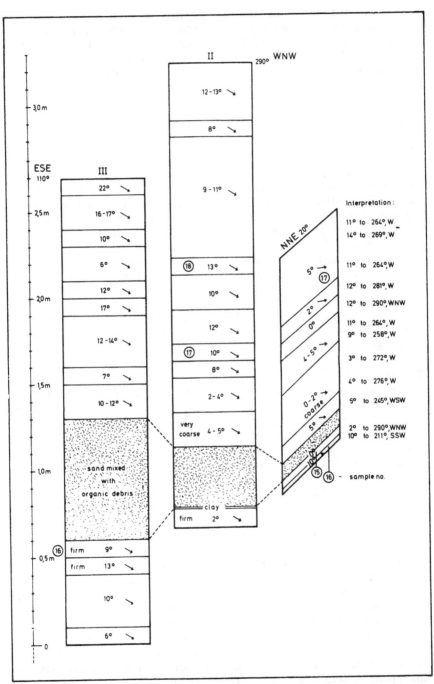

Figure 3 Plan of the two dune cuts at right angles to each other with three profiles showing bedding dips and directions and sampling sites. (True values calculated to the right).

are in accordance with modern prevailing winds. The sand is compacted, rather coarse-grained, and partly topped by thin silt and clay layers, followed by a sand mass without any bedding structures, but containing clay lenses with desiccation cracks approximately 35 cm below its upper boundary. In this upper part, the sand is thoroughly mixed with plant debris, snail-shells and clay flakes, and in some places, also contains root casts and carcasses of locusts and *Tenebrionide* beetles. Plant debris were collected for dating by the radio-carbon method and yielded a ^{14}C age of 1835 ± 35 years B.P. (Lab. Heidelberg). Extremely coarse-grained sands occur on top of this "bioturbation" horizon. Towards the top of the profiles the grain size generally decreases, except for thin isolated bands of very coarse sand. The strata dip in westerly directions as they do beneath the surface of the western flank of the present dune. The angles of dip vary between 9 and 14°, comparable to the eastern flank of the modern dune.

Sand strata beneath the 20° inclined modern western surface, however, are not really parallel to the surface, but dip at reduced angles of about 13° (evidence from test cuts for comparison). This is surprising because the profile of the Toshka-Canal dune is very similar to the seif dune studied in detail by Mckee and Tibbitts (1964) in Libya. Here, the dip was the same or steeper than the surface on both slopes. According to Mckee and Tibbitts (1964), beddings at low angles <15° are formed by accretion.

Textural analysis

During the investigation of the dune interior, four sand samples (nos. 15-18) were taken from distinct horizons for textural analyses. The sands were sieved using a ϕ-set and their grain size parameters were calculated after Folk and Ward (1957). For comparison, blown sand from the vicinity (sample no. 2) was collected and dealt with similarly.

From previous dune studies in various deserts (Besler 1980, 1982a, 1982b, 1984) it has been found that the frequency distributions of grain size in sands are especially promising for the interpretation of depositional dynamics (Walger 1964). The frequency distributions of grain size in sands from the interior of the Toshka-Canal dune are presented in Figure 4. The data have been plotted using the technique of Walger 1964. In these diagrams, weight-percentage divided by grain size interval (%/mm) is plotted on the Y-axis. This type of graph has been shown to be useful in the interpretation of depositional dynamics (Besler 1984). The resultant graphs are very characteristic and informative if compared to the many curves already drawn of sands from other dunes and deserts.

The bottom sample (number 16) is very well sorted and displays a rather symmetrical frequency distribution with a clear maximum in the fraction 0.25 - 0.5 mm. There is no similarity to the modern blown sand (no. 2 in Fig. 5) which is more fine-grained. It should be mentioned that the actual blown sand in most modern deserts,

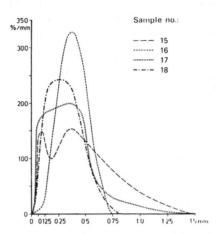

Figure 4 Frequency distributions of grain size in sand samples from the interior of the Toshka-Canal dune. Note the dimension on the Y-axis.

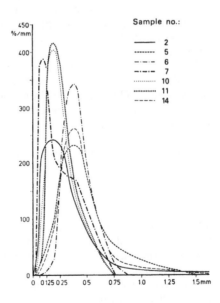

Figure 5 Frequency distributions of grain size in sands sampled along the profile across the Toshka-Canal dune, No. 2, for comparison represents blown sand from the vicinity. Note the dimension on the Y-axis.

with a few marginal exceptions, has its maximum in the fraction 0.125 - 0.25 mm. There is a certain similarity in sample 16 to the curve of the modern surface sand number 6 (see Fig. 5) from the lower western slope, but the skewness is negative in the latter case, indicating deflation of the finer grains. The sample 16 sand is comparable, however, to barchan-dune sands from the Kharga-Dakhla region in Egypt (Besler 1983), and from the Algerian Sahara (Besler 1984), as well as other places. Barchan-dune sands generally have a broad and rather symmetrical maximum without tails in the coarser and/or finer fractions.

The assumption that barchan sands exist on the bottom of the Toshka-Canal dune is corroborated by sample number 15 from layers immediately below the organic horizon. The biomodal frequency distribution is very typical of sands from sand shields. These features are comparable to McKee's (1966, 1982) dome-shaped dunes, but the term shield is preferred here as these sand bodies are very flat and do not have a dome-like appearance. Sand shields investigated in the Namib Desert and in the New Valley region (Besler 1983) possess great textural similarities, all of them showing a deflation minimum in the 0.125 - 0.25 mm size fraction. Sand shields are stabilized sand accumulations which act as transportation surfaces; they may develop into barchans (Embabi 1970/71), or barchans may degenerate into sand shields. In the New Valley, all stages in these transitions can be studied (Besler 1983). It seems that in the Khor Toshka, barchans degenerated into sand shields during increasingly humid conditions, and later even became vegetated.

The sands on top of the organic horizon are completely different. Their mean grain size is smaller, and sorting is only moderate. The latter is expressed by a flattened maximum and a tail of coarse grains. Sample number 17 still represents a transitional stage between the bimodal shield sand and the actual lee-dune sand (compare Fig. 5). Sample number 18 reflects progressing aeolian selection and better sorting, and approaches the frequency distribution of numbers 10 and 11, the crest and crestal slip-face sands of the modern dune.

Interpretation

Bedding dips and textural characteristics of the sand at the bottom of the Toshka-Canal dune indicate that small barchan dunes were the initial precursors. According to structural studies by McKee (1966), barchans should display foresets at steeper bedding angles. On the other hand, his cross section of a barchan dune (McKee 1966, Fig. 8) shows large portions having bedding dips comparable to the Toshka example. Moreover, the barchan dune may have become flattened into a sand shield during development. At White Sands National Monument, dome dunes were found to be bevelled barchans with steep beddings, but similar features investigated in the Namib Desert (McKee 1982, Fig. 37) also had strata dipping at 11-13°.

Winds were prevailing from the same directions as today (NNE). The moving barchans then degenerated into immovable sand shields stabilized by moisture. On several occasions, the sands were partly covered by standing water as evidenced by clay lenses (up to 2 cm thick) with desiccation cracks. At this time, there may have existed a rolling sand surface which is preserved only beneath the modern dune, and is eroded by wind in the vicinity. Almost 2000 years ago, the sands were probably completely covered by vegetation providing a favourable environment for insects, beetles and snails. Both clay bands and organic remains were also found in other dunes of the New Valley, indicating perhaps a period of less extreme aridity than today. Later, the lee dune growing downwind from the northern scarp reached this locality, incorporated the older sandy deposits and protected them against erosion. The rather low dip of strata and the isolated layers of very coarse grains indicate accretion processes alternating with deflation.

Dynamics of the modern dune

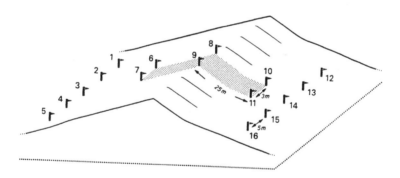

Figure 6 Arrangement of blue sand and of markers for measurements on the southern part of the Toshka-Canal dune.

Dynamic analysis
In order to gain knowledge about the present movements of sand on and along the Toshka-Canal dune, the site was prepared for a study of dynamics one month before the actual field work was undertaken. A strip of sand across the dune, 3 m wide and 2 cm deep, was dyed blue February 16, 1982 (Fig. 6). Previous experiments in the Namib dune field showed that spraying with Methylene Blue dissolved in water can be used with good results; quartz grains are easily and intensely coloured, the grains will not stick together to form aggregates, and even several days of rain

will not wash out the blue colour. Sixteen markers were placed on the dune: two on the crest, two at each end of the blue strip on the flanks (3 m apart) and five (5 m apart) in prolongation of the blue strip at the base of each flank (Fig. 6). The sand movements, reflected by the changes in sand height at these markers, were measured daily until March 11, 1982.

Immediately after spraying, a photo documentation began and was carried on until there were no longer any perceptible changes in colouring (February 27). Coloured photographs were taken daily from the same four positions on the crest and at the base always looking in the same directions. Unfortunately, the first film was destroyed, but the remaining films showed the gradual change in sand colours. For final observations, diggings were made on March 11, 1982, along what had been the blue strip. During the weeks of observation, northerly winds were predominant with mean velocities of 16 km/h measured at the nearest weather station of Abu Simbel, but calm days and a sandstorm were also recorded. Unfortunately, no detailed daily wind records for the site were available from Abu Simbel, but wind directions and strength were noted on the site twice a day by the engineer in charge.

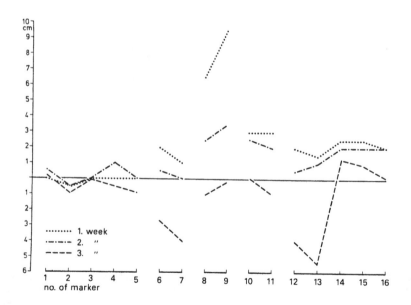

Figure 7 Changes in sand height at the sixteen markers on the Toshka-Canal dune, measured during three weeks with predominant northerly winds in February-March 1982.

Figure 7 shows the sand movement in centimeters for the 16 markers, totalled weekly. The zero-line represents the dune surface at the day of dyeing. During the first week after colouring (February 16-23, 1982), winds were rather calm. Nevertheless, sand heights increased with the exception of the western base. Since the process of dyeing results in a disturbance of natural conditions, some of the sand movement was due to a re-arrangement of the equilibrium. Activity was greatest in the crest region where the dune was growing at its maximum rate (9.5 cm). This can be explained partly by the fact that sands in the crestal area are finer than elsewhere (see textural analysis). The blue sand had already vanished completely by February 22. On the middle section of the western slope, the blue surface was the first to be covered by sands, and on February 22, was excavated again during the occurrence of westerly winds. Sands carried by the subsequent northerly winds covered the blue surface again. On the middle section of the eastern slope, the blue surface was only partially covered. Here, sand transport parallel to the dune flank was documented by a blue veil extending south of the dyed strip. Accumulation occurred at the eastern base of the dune. On the whole, the altered sand surface with more accumulation on the eastern dune flank seems to reflect the exceptional day with westerly winds.

During the second week (February 24 - March 2, 1982) moderate northerly winds were recorded, and the crest height decreased by 6 cm while on the middle section of the western slope, sand heights decreased to almost their initial position and blue sand was still visible. Photographs revealed that during rather calm days or days with moderate northerly winds, the blue areas were covered, and during exceptional days with southerly winds, these cover sands were removed again. On the eastern slope there was only a slight decrease in dune height and the blueish tint extended farther to the south. During the days with southerly winds, these areas were covered by non-coloured sand. Obviously, southerly winds were deflating sand on the western flank and accumulating sand on the eastern flank.

During the third week (March 3-10, 1982) with strong northerly winds, great changes occurred. The crest height was reduced to its initial position or lower and remnants of blue sand again became visible. On the middle section of the western slope, there was a large decrease of sand, but in some places a thin layer of 1-2 mm blue sand, which had been covered previously, still remained at the end of the week. No traces of blue sand were found in diggings on the upper western slope, but on the lower western slope, the blue sand was still buried in places by as much as 10 cm of non-coloured sand. The western base seemed stable during the entire time. On the middle section of the eastern slope, deflation reduced the sand height to its initial position as on the crest. Here, only slight traces of blue sand were found in the final diggings.   At the eastern base, the sand budget was locally different, reflecting perhaps a slow shift of the basal undulations in the dune ridge.

It is remarkable that the dune height increased during weak northerly winds and was reduced again during strong northerly winds.   The reduction seemed to be greater on the western slope according to its more upwind position in the SSW-trending dune ridge.   For the same reason, the crestal slip face was pointing to the east.

Textural analysis

During the final observations, six sand samples were taken along the dune profile (Fig. 8).   Their grain size characteristics were calculated and their frequency distributions drawn (Fig. 5).   Sample numbers 10 and 11 from the crest and the base of the small crestal slip face towards the east have almost identical frequency distributions and grain size parameters.   This similarity between crestal sands and slip-face sands has been found in many dunes of other deserts (Besler 1977, 1980, 1984, in press).   This is due to the fact that gravity, and not wind, is the dominant agent of transport from the crest to the slip face.   Narrow, high peaked curves in the grain-size class 0.125 - 0.25 mm,   the actual blown-sand fraction, without basal tails are characteristics associated with the mobility of these sands.   They are well sorted, positively skewed and mesokurtic.   In the response diagram (Besler 1983) which reflects the response of sands to actual aeolian activity, they are categorized under the section of aeolian mobility.

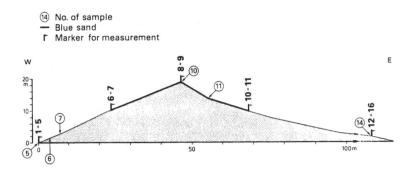

Figure 8 Profile at true angles across the southern part of the Toshka-Canal dune, showing positions of markers and sampling sites.

Another group of curves represent sample numbers 5, 6 and 14 from the base of both dune flanks. These sands are much coarser with a maximum occurring in the fraction 0.25 - 0.5 mm and have even coarser tails (up to 1.0 mm). They are only moderately sorted and leptokurtic, both properties reflecting little response to actual winds. Sands from the western base (nos. 5, 6) are negatively skewed while sands from the eastern base (no. 14) are positively skewed. Since negative skewness indicates a lack of fines, this is an additional indication that deflation is favoured on the western slope. Sample number 6, however, with a higher peak and a nearly symmetrical base approaching the characteristic sorting of barchan sands, is an exception. In addition, the sand is very well sorted and mesokurtic, indicating mobility. Perhaps another small barchan dune has been incorporated here into the bottom portion of the lee-dune ridge. In the response diagram, all these sands are categorized under the section of aeolian residuals, a sub-section of aeolian stability (Besler 1983), but this diagram, according to testing with hundreds of sand samples, is not valid for barchan dune sands.

Sample number 7 taken from approximately 7 m upslope on the western flank shows a peculiar frequency distribution with a high and narrow peak in the fraction 0.063 - 0.125 mm, and a broad shoulder at 0.25 - 0.5 mm. The sand is poorly sorted and very platykurtic. Strong platykurtosis is characteristic for poorly mixed sands from different sources not yet in equilibrium with wind activity (Besler 1980). This sand is also categorized under aeolian stability in the response diagram. The fine grains are sheltered by the coarse grains forming a kind of desert pavement. The sand could be a mixture of basal dune sand (0.25 - 0.5 mm) and actual blown sand from the surrounding surface which has a fairly high percentage of grains in the 0.063 - 0.125 mm fraction. Fine grains may have been deflated from the base at numbers 5 and 6 (negatively skewed) and blown upslope, but not in the necessary quantity to produce the frequency distribution of number 7.

Interpretation
According to analytical evidence (dynamic and textural) only the crestal section of the dune, i.e. the upper 10 m of both slopes including the slip face, is active. Here, the sands are mobile and respond to wind activity as is evidenced by the shifting of the peaks of the cock's comb structure.

On the dune flanks, the sand is distinctly coarser with little variation. During the dominant moderate northerly winds, the western slope is an accretion surface (more upwind) with the eastern slope being a transportation surface; during strong northerly winds, deflation occurs on both. The same occurs during rare southerly or westerly winds, the former causing accretion on the eastern slope (more upwind), but not to the same extent as accretion on the western slope because southerly winds are not as dominant. This is also expressed by the dune profile.

The basal sections are stable with the sands being much coarser (mean grain size around 0.4 mm), and showing no response to normal wind activity. They are residuals of an advanced stage as the fine-sand fractions (0.063 - 0.25 mm) are almost totally removed. However, at the eastern dune base there appear to be slow shifts of the zibar-like undulations during sandstorms. This would account for the fact that the very fine grains, usually sheltered in deflated sands, are absent and therefore the sand is not bimodal.

Sand is transported south via the bulk movements of the crestal section and as blown sand along the surfaces of the middle flanks. Thus, the dune ridge slowly grows towards the south.

Tachymetric and air photo survey
A tachymetric survey of the southern end of the northern section of the Toshka-Canal dune was made on December 12, 1981 and again on March 11, 1982. Within these three months, a remarkable change in topography occurred. During 1974/75, the northern dune ridge was crossed by vehicles close to what is now its southern end. As a consequence of this disturbance, a shield dune developed following the dissection. A shield dune is a transitional stage between a sand shield and a barchan dune (Embabi 1970/71), and can be described as a round or elongate sand body with a small marginal slip face, but lacking horns. In December 1981, the dome-shaped sand body was 60 m long and 5.7 m high with the slip face being only 3.1 m high and situated at the margin 25 m south of the highest point. In March 1982, the slip face was already 4.2 m high and only 8.5 m south of the highest point. As well, the sand body was now much broader (71 m) than it was long (30 m), the ground plan being almost a semi-circle, yet, there were no distinct horns. Broadening of the sand body and vertical as well as lateral growth of the active slip face account for the transformation from a sand shield to a true barchan dune.

At the northern end of the southern section of the Toshka-Canal dune, a wind gap developed within three years after the dissection. The higher part of the dune ridge north of this wind gap acquired a barchan-like shape in 1981. In March 1982, the southern dune ridge began as a giant barchan, 14 m high, with a slip face descending 8 m deep into the wind gap. Obviously, after the dissection the wind flow was no longer directed along the sand body, but met the dune ridge more obliquely causing a wind gap. The separated sand body then developed rather quickly (within one year) into a barchan riding on the ridge. Long rows of barchans riding on each other in a shingle structure are known from the Rub' at Khali (Besler 1982) and other deserts. They may have developed similarly from longitudinal dune ridges under obliquely blowing winds, causing wind gaps at certain intervals.

In the vicinity of the Toshka-Canal dune there are several, mostly smaller, lee-dune ridges of the same type. Since these dunes are still undisturbed by human activity, their natural growth or downwind prolongations were measured to allow for comparison with a map based on aerial photographs taken during 1963. These dunes show great variation in their rates of movement. The longest (1.6 km) and the shortest dune (0.3 km) had grown 4-5 m/year; the middle-sized dune (1.1 km), however, had grown 22.5 m/year. The latter was found to be not entirely undisturbed, as was evidenced by the many vehicle tracks found at the former southern end of the dune, i.e. approximately 400 m north of the present southern end. At this site, the lee-dune ridge with a sharp crest line changed into several sand shields declining in height. This transformation shows similarity when compared to the southern end of the northern section of the Toshka-Canal dune. Perhaps the abnormally high growth rate of this dune was due to barchan breeding triggered by vehicle crossings in the manner described above. Twenty m/year were found to be the rate of migration for small barchan dunes farther north in the Kharga Depression (Müller 1981). The barchans may have degenerated into sand shields again for unknown reasons. It should be noted that no barchans are found in the vicinity of the Toshka-Canal dune although there is much blown sand present.

Discussion and conclusions

According to the above investigation, which used various analytical methods and techniques, it is possible to understand the formation and dynamics of the Toshka-Canal dune. This longitudinal dune ridge was growing as a lee dune behind a prominent part of the northern bank of the Khor Toshka when conditions became drier again following a moister period. Comparing the growth rates of undisturbed lee dunes in the vicinity and assuming comparable average wind velocities during the historical past, the age of the Toshka-Canal dune would be some hundreds of years. Taking into account the fact that this special dune, unlike the other lee dunes, was prohibited from progressing any further by the southern bank of the Khor, the age may even be a thousand years.

At least 1835 years B.P., the Khor Toshka was locally vegetated and contained pools of standing water. These conditions, however, prevailed for only a short time following another period of sand movement in the form of small barchan dunes. These older sandy deposits were preserved beneath the lee-dune and have been locally eroded by wind. Although the organic matter found in the dune interior would normally provide the rare chance for dating the dune formation, this cannot be done easily because several possibilities have to be considered:

(a) Around 1835 years B.P. the area became drier, but it took another thousand years before the Toshka-Canal dune began to grow.

(b) The lee dune, formed by sand supplied from the northern plateau, was overrunning the vegetation in the Khor Toshka around 1835 years BP and was slowed down by these local conditions.

(c) The lee dune has been in its present position, ending at the south bank, since approximately 1835 BP.

(d) The Toshka-Canal dune, being older than comparable lee dunes, was formed by weaker winds in its early stages. However, there is no evidence for this possibility from grain size characteristics.

(e) The growth rates of undisturbed lee dunes in general decrease with increasing dune length and dune height because more of the sand blown along the dune body is trapped.

The last possibility seems to be rather plausible. The only certainty is, however, that the Toshka-Canal dune is younger than 1835 ± 35 years BP. The dynamics of this lee dune are not comparable to seif dunes where diagonally blowing alternating winds cause oscillating slip faces and sand encroachment on both sides as is evidenced by cross bedding. In the case of lee dunes, the primary wind flow is mainly parallel to the dune ridge with deflation, sand transport and accretion alternating on both flanks according to different wind velocities. Accretion is caused by moderate winds, whereas strong winds lead to deflation and downwind extension. The investigated dune ridge runs SSW due to prevailing NNE winds, but rare winds from other directions also model the sand body. Westerly winds seem to be especially efficient; they meet the dune ridge obliquely as occurs in a true seif dune, and therefore cause accretion on the western and encroachment on the upper eastern slope. Because of the swinging crest line, the lee currents are differentiated and cause undulations on the lower eastern slope (almost absent on the western slope). But these are only modifying effects. The ground plan of the dune, as in other lee dunes, is very straight and does not curve as in seif dunes.

A different wind flow pattern develops if the parallel flow along the dune ridge is disturbed, for instance by dissection or even by a small incision caused by vehicle crossings. If the disturbances occur close to the downwind end of the dune ridge, the separated sand body develops a slip face and slowly grows into a barchan dune which may migrate. If an incision is made in the middle of the dune ridge, a wind gap develops, and the thereby separated sand body similarly grows into a barchan dune, in this case riding on the main dune ridge. These patterns may develop within only a few years.

Acknowledgements

I thank Prof. D. Busche, University of Wuerzburg, for his cooperation in the field. The High Dam Authority at Aswan provided assistance and equipment.

References

Besler, H. 1977. Fluviale und aölische Formung zwischen Schott und Erg. In *Geographische Untersuchungen am Nordrand der tunesischen Sahara*, W. Meckelein (ed.), 19-81. Stuttgarter Geographische Studien 91.

Besler. H. 1980. Die Dünen-Namib: Entstehung und Dynamik eines Ergs. *Stuttgarter Geographische Studien* 96.

Besler, H. 1982a. A contribution to the aeolian history of the Tanezrouft. *Bulletin de l'Association de Geographie Français* 84, 55-9.

Besler, H. 1982b. The north eastern Rub' al Kháli within the borders of the United Arab Emirates. *Zeitschrift für Geomorphologie N.F.* 26, 495-504.

Besler, H. 1983. The response diagram: distinction between aeolian mobility and stability of sands and aeolian residuals by grain size parameters. *Zeitschrift für Geomorphologie N.F., Supplementband* 45, 287-301.

Besler, H. 1984. Verschiedene Typen von Reg, Dünen und kleinen Ergs in der algerischen Sahara. *Erde* 115, 47-79.

Besler, H. In press. Dunenstudien am Nordrand des GroBen Østlichen Erg in Tunesien. *Stuttgarter Geographische Studien* 100.

Embabi, N.S. 1970/71. Structures of barchan dunes at the Kharga Oases Depression, the Western Desert, Egypt (and a comparison with structures of two aeolian microforms from Saudi Arabia). *Bulletin de la Société de Géographie d'Egypte.* 43/44, 53-71.

Folk, R.L. and W.C. Ward 1957. Brazos River Bar: a study on the significance of grain size parameters. *Journal of Sedimentary Petrology* 27, 3-26.

Haynes, C.V. 1980. Geochronology of Wadi Tushka: Lost tributary of the Nile. *Science* 210, 68-71.

McKee, E.D. 1966. Structures of dunes at White Sands National Monument, New Mexico (and a comparison with structures of dunes from other selected areas). *Sedimentology* 7, 1-69.

McKee, E.D. 1982. Sedimentary structures in dunes of the Namib

Desert, Southwest Africa. *Geological Society of America, Special Paper* 188, 1-64.

McKee, E.D. and G.C. Tibbitts 1964. Primary structures of a seif dune and associated deposits in Libya. *Journal of Sedimentary Petrology* 34, 5-17.

Müller, H.D. 1981. *Die Entwicklung des Oasengebietes von Al-Kharga (Ägypten) als Resultat økologischer Bedingungen und sozioøkonomischen Wandels.* Diplomarb. Gøttingen.

Walger, E. 1964. Zur Darstellung von

KorngrøBenverteilungen. Geologische Rundschau 54, 976-1002.

8

Aeolian processes and dune characteristics of a developed shoreline: Westhampton Beach, New York

Karl F. Nordstrom, James M. McCluskey and Peter S. Rosen

Abstract

Aeolian processes and dune forms are examined on an eroding barrier island using sand trap data, a calculated aeolian sediment budget, and field observations of dune characteristics. Results indicate that the highest rate of aeolian transport occurs on the beach. Transport parallel to the shoreline greatly exceeds transport onshore and offshore. The calculated annual net rate of aeolian transport varies from 0.31 m^3/yr/m of ground surface to 0.61 m^3/yr/m in developed areas and averages 0.74 m^3/yr/m in nearby natural areas. The direction of net aeolian transfer ranges from east (alongshore) to northeast (onshore). Wave erosion of the base of the dune, followed by aeolian transport from the beach is the primary mechanism for landward displacement of the dune. Residents place sand fences parallel to the shoreline to maintain a dune between the ocean and the first row of houses. Wave erosion results in a sequence of changes in dune form, progressing from a low, wide feature to a narrow "knife-edge" shape. The interaction of wind and sand transport with houses, combined with landscaping efforts of shorefront residents, creates a crenulate backdune. Eventually, the beach narrows, through erosion, to the point where a dune cannot survive storm wave attack. Aeolian processes then play a minor role in dune formation until artificial beach nourishment establishes conditions for a new cycle to begin. Groins alter the rate at which dunes pass through sequential stages of development, and they aid in establishing an irregular shaped dune in the protected shoreline.

Purpose

Coastal dunes provide protection to shorefront buildings from storm flooding and wave attack. Planners, managers, geomorphologists, and engineers have long been aware of the importance of dunes, particularly on low lying barrier islands. Previous studies have focussed on management issues such as identifying ways to prevent human degradation of the dunes (Godfrey *et al.* 1978, Baccus & Horton 1980, Leatherman 1980, Nordstrom & McCluskey 1984) or on engineering applications such as identifying the rates of erosion of coastal dunes (Edelman 1972, Van de Graaf 1977), the levels of protection dunes provide (Gares *et al.* 1980), or the best means of creating a dune using sand fences or vegetation plantings (Savage & Woodhouse 1968, Knutson 1978, Snyder & Pinet 1981, CERC 1977).

There is a considerable difference between the characteristics of the optimally designed protective dune and the dunes which exist in developed communities. Part of the explanation for the difference is that the information generated by coastal scientists is rarely used by shorefront residents to manage their properties. Many dune-building programs are conducted on an ad-hoc basis, and variations in local practices result in a variety of dune configurations. The form of the dunes is also affected by buildings and shore protection structures which cause interruptions to wind flow and patterns of sedimentation.

Recent studies of developed barrier island communities have provided information on the rates of migration of foredunes (Gares 1983) and the effects of buildings on dune forms (Nordstrom & McCluskey 1985). The purpose of this study is to complement previous investigations by providing quantitative estimates of the effect of aeolian processes in this migration and identifying the changes in the form of the dunes as they are displaced inland. Westhampton Beach, New York (Fig. 1) was selected for detailed analysis because it has a long history of development, and it has been intensively studied in the past. Shoreline erosion has been critical at Westhampton Beach, and several shore protection projects have been implemented. Regional differences in these human adjustments to erosion provide a variety of dune conditions along the shore which facilitates comparative analysis.

Aeolian processes and dune forms are analyzed using: 1) measurements of dune dimensions derived from topographic maps; 2) an aeolian sediment budget calculated using sand trap data and local wind records; and 3) field observations of dune characteristics in the vicinity of structures. This information is then used to describe dunes in terms of stage of development and position relative to houses and groins. Also included is an assessment of the effects of selected shore protection alternatives on aeolian transport.

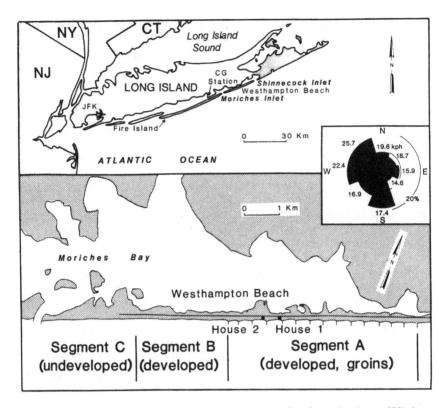

Figure 1 Study area, Westhampton Beach, NY. Wind rose data are from Kennedy Airport (JFK) for a five year period 1975-79.

Study area

Westhampton Beach is a 24 km long barrier island located on the south shore of Long Island (Fig. 1). Tides at Westhampton Beach are semi-diurnal with a mean range of 0.9 m and a spring range of 1.1 m. The average height of breaking waves is 0.73 m; mean breaker period is 8.1 seconds (DeWall 1979). The direction of net wave-induced longshore sand transport (drift) is from east to west. Dominant winds are from the northwest and west, and the direction of net sand transport by winds is from west to east. The south shore of Long Island is one of the few coastal locations on the east coast where the net movement of sediment by winds is opposite the direction of net movement by waves.

The ocean beaches consist of quartz sand with median diameters ranging from 0.3 to about 1.4 mm (Taney 1961, Snyder & Pinet 1981) and are approximately 75 to 125 m wide from the frontal dune to the mean low water shoreline (DeWall 1979). Dune sediments have an average median diameter of 0.41 mm (McCluskey *et al.* 1983). Maximum dune elevations range from 4.6 to 7.6 m above mean low water (DeWall 1979).

Beach erosion at Westhampton Beach has increased dramatically in the past 50 years. The annual rate of shoreline retreat from 1838 to 1933 was less than 0.5 m/yr. From 1940 to 1956 the rate of retreat was 2.5 m/yr (Taney 1961). Accelerated shoreline erosion has been attributed to the opening of Moriches Inlet in 1931 and Shinnecock Inlet in 1938 (Fig. 1) and to the artificial stabilization of these inlets in 1952 (McCormick 1973). Shore protection projects were implemented as early as 1938. Past projects include two major beach fill operations totaling over 2.7 million m^3, three separate small scale dune fill projects totaling over 192 000 m^3, emergency closures of breaches in the barrier following storms in 1954 and 1962, and construction of fifteen groins between 1965 and 1970 (DeWall 1979). Sand fences and vegetation plantings have been used over the years to build protective dunes.

The most significant management problem in the study area is the severe wave erosion which has occurred west of the terminal groin (Fig. 2). Many houses have already been destroyed, and many others are in imminent danger. The erosion problem has been exacerbated by the failure to complete the groin field along the full length of developed shoreline as called for in the Corps of Engineer plan of protection (Long Island Regional Planning Board 1984). The failure to build all of the groins and fill them with sand is attributed to the desire of the local interests to reduce project costs (Heikoff 1976). Differences in the beach sediment budget in protected and unprotected segments have resulted in different dune characteristics in these areas.

Figure 2 Erosion problem area in Segment B looking north, March 1980. Terminal groin is just to east (right) of photo.

Three segments of the island have been selected for detailed analysis (Fig. 1). These include: (a) the 5.5 km long shoreline developed with houses and protected by groins (Segment A); (b) the 2.3 km long shoreline developed with houses immediately downdrift of the groins (Segment B); and (c) the 2.6 km long public park (Segment C), which is undeveloped except for a parking lot and bath house in the eastern portion of the segment. Dunes are wider and higher in Segment A, which is protected by groins (Table 1), than in the sediment starved segment downdrift (Segment B). In Segment C, which is farthest downdrift from the groins, sediment starvation is reflected in narrow beach widths, but shoreline retreat has not proceeded to the point where dunes have been eroded. Dunes are higher in Segment A than in Segment C because the residents have built the dunes up using sand fences and vegetation. The fences are usually placed on the seaward slope of the dune in an attempt to extend the dune farther seaward, but they are also placed on top of existing dunes and at low points in the crest line to increase dune elevation. Fences are usually placed parallel to the existing dune line in a straight or zig-zag configuration.

Table 1 Characteristics of beaches and dunes at Westhampton Beach study area. (Data from USACOE 1980).

Location	Average dune width (m)	Average maximum dune height (m above MSL)	Average distance, beach berm crest to dune crest (m)*
At 100m transects (N=108)			
Segment A (N=56)	61.6	6.2	66.5
Segment B (N=25)	15.6	4.8	36.9
Segment C (N=27)	61.9	5.2	38.5

*The width of the beach changes daily; these figures are valuable for the insight they provide on relative differences among reaches, rather than for the absolute values.

Determination of rates of eolian transport

An aeolian sediment budget was determined for the study area using rates of transport gathered in the field. Field data were obtained from 36 aeolian sediment samplers of the Rosen (1979b) design emplaced at two representative house sites at Westhampton Beach (Fig. 1). Both houses are located landward (north) of a vegetated dune crest on a partially vegetated surface composed of wind blown sand. The median diameter of three surface samples taken at each house is 0.36 mm at House 1 and 0.41 mm at House 2. The houses are part of a single row of detached dwellings located between the beach and the access road. There are additional houses located north of the road. The distance between shorefront houses averages about 7 m.

Six arrays of sediment samplers were emplaced to monitor the movement of sediment in the vicinity of each of the structures. The traps were placed on the backbeach, dune crest, and backdune, in arrays of four, with one trap oriented toward each of four components of the shoreline transport system (onshore, offshore, and each direction alongshore); bi-directional (onshore-offshore) arrays were used at locations on the east, west, and north of the buildings. The traps placed to the rear of the dune were directly offshore of the houses; the traps placed to the north were directly onshore of the houses. One hundred thirty eight additional traps were emplaced in undeveloped portions of Fire Island to characterize sediment transport in undeveloped dunes (McCluskey *et al.* 1983).

The traps at Westhampton Beach were in place March 23 to April 20, 1982. Measurements were taken daily except for five days when snow was on the ground and traps were clogged. Rates of sediment transport which occurred during the sampling interval at the two houses are presented in Table 2. These rates are representative of only the wind conditions which occurred during the field study. The data were used to calculate a more representative annual net sediment budget, presented as a vector of sediment transport, using meteorological data for 1980 from the Moriches Coast Guard Station (Fig. 1) and a modification of the Kadib (1964) calculation procedure (described in McCluskey *et al.*, 1983).

Table 2 Volume of sediment trapped in field samplers 3/23/ − 4/20/82.

Trap Site Station Name	Total volume of sediment trapped in 10^3 x cm^3/linear m			
	Onshore	Offshore	West Alongshore	East Alongshore
House I				
Beach	161.7	27.2	213.6	31.7
Dune Crest	11.8	2.3	2.0	8.1
Rear of Dune	4.1	5.2	2.2	2.8
East Side of House	9.3	15.0	N/A	N/A
West Side of House	6.0	18.6	N/A	N/A
North Side of House	3.0	1.4	N/A	N/A
House 2				
Beach	115.2	42.4	150.1	86.6
Dune Crest	12.2	3.3	4.2	5.7
Base of Dune	0.5	2.0	1.2	0.8
East Side of House	1.7	1.9	N/A	N/A
West Side of House	1.9	2.1	N/A	N/A
North Side of House	5.6	1.7	N/A	N/A

Discussion of transport rates

Data gathered in the field (Table 2) reveal that the highest rates of aeolian transport are on the beach. This occurs because of the plentiful amount of sand and the absence of vertical obstructions to the wind. The rate of offshore transport on the beach and at the dune crest is considerably less than the rate of onshore transport because of the reduction in wind velocity and transported sediment caused by vegetation on the backdune. Between the dune crest and the landward side of the house, the net transport is offshore, whereas the net transport on the seaward side of the dune crest is onshore. These opposing flows contribute to the creation of a high dune.

The calculated annual rate of net aeolian transport in the vicinity of House 1 is 0.61 m³/yr/m. Net movement is toward the east-northeast, nearly parallel to the shoreline. The net transport at House 2 is 0.31 m³/yr/m toward the north-northeast at an angle of 43° to the trend of the shoreline. The calculated net rate of transport at natural areas on Fire Island is similar to these values, with an average of 0.74 m³/yr/m moving alongshore toward the east.

Although annual net sediment transport is to the east, there may be periods of pronounced westerly transport (Table 2). The important consideration is that the amount of transport parallel to the shoreline greatly exceeds the transport onshore and offshore. Similar findings are reported on undeveloped barrier islands where dominant winds are more perpendicular to shoreline trend (Rosen 1979a). The implications of these findings are that human structures which are built perpendicular to the shoreline will have a pronounced effect on the location of aeolian sedimentation. These structures will trap sand moving in both directions alongshore, and the volumes retained by them will be a function of the gross transport rate rather than the net rate.

The rate of transport alongshore by aeolian processes is considerably less than the rate of wave-induced longshore transport. Using an average width of beach determined from field surveys conducted at Westhampton (DeWall 1979), the calculated net rate of aeolian transport along the entire beach fronting House 1 is 57.4 m³/yr. The sediment budget for wave-induced longshore transport past a given location in Segment A is 154 125 m³/yr, toward the west (Research Planning Institute 1983).

The important consideration in terms of dune maintenance is the shore-normal expression of the sediment budget. A negative beach sediment budget results in the onshore displacement of the shoreline profile through time. The position of coastal dunes is determined by the location of the beach, and these two components of the shoreline system migrate together.

Changes in dune characteristics through time

Managed and vegetated coastal foredunes are not stationary. They move inland through onshore aeolian transport over the top of the dune crest, and they are displaced inland through wave erosion of the seaward portion of the dune followed by wind deposition landward of the former position of the beach/dune contact. Wave-induced changes result in the most rapid onshore movement. The rate of aeolian transport from the beach to the base of the dune can be over ten times greater than the rate of transport over the crest (Table 2).

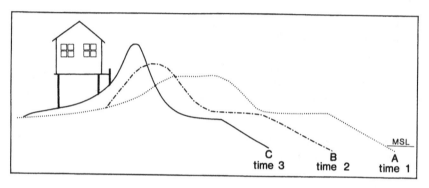

Figure 3 Model of changes on the beach and dune through time when dune position is stabilized by shorefront landwoners.

Attempts to maintain the position of the dune crest seaward of houses is made through the use of sand fences. Typically, the fences are placed on the seaward side of the existing dune, which creates a broad dune platform (Fig. 3, profile A). Sand fences at Westhampton Beach will fill to capacity within a year (Snyder & Pinet 1981). The volume of sand trapped by a single fence can vary from 5.9 m^3/m of ground surface to 8.4 m^3/m (CERC 1977). This volume is sufficient to allow new dunes to form between large, low-frequency storms, providing that there is sufficient beach width to protect the dune against wave attack from small, high-frequency storms. As sediment is removed from the beach/dune system through wave erosion, the zone in which dunes can form narrows. A narrow dune must be higher to contain the same volume of material as a wider dune. The increased crest height is required to provide protection against wave runup that the narrow beach can no longer provide (Fig. 3, profile C). High dunes are perceived by many residents to provide better protection against storm wave attack than lower dunes although the low dunes may be located farther seaward and contain a greater volume of sand. The high dunes are less effective as a means of protection because they are unstable due to the narrow width and steep slope. They are susceptible to crest lowering through trampling, and it is easier to create notches in the crest which favor formation of blowouts. Onshore sediment transfers are greatly accelerated through these notches. Results of a 43 day field study at Fire Island (McCluskey *et al.* 1983) show that a 3 m wide by 1.5 m deep notch in a dune crest 3.4 m above the back beach resulted in a rate of transport 418% greater than across an equivalent length of dune crest 20 m away.

The dune crest in the immediate vicinity of houses retains the linear appearance of an engineered dune on the oceanside and develops a crenulate, barchanoid shape on the landward side (Fig. 4). The barchanoid shape is the result of both the interference with the wind stream by houses and the direct attempts by residents to modify the location of sediment deposition using earth moving equipment or vertical walls. The deceleration of onshore wind in front of the vertical obstructions caused by houses favors echo dunes (Tsoar 1984). Offshore winds are accelerated as they blow around the house; the barchanoid shape contributes to convergence of the wind stream which scours the face of the bay of the dune (Fig. 4). Residents often erect bulkheads at the base of buildings to prevent sand inundation of decks and patios by devegetated backdunes. These walls restrict the active zone of aeolian activity to the upper portion of the dune.

Figure 4 Crenulate shaped bay of dune at house site in Segment A showing roots of vegetation (far left) exposed by scour from offshore winds and attempts of resident to prevent sand inundation at base of structure.

The size of the horns of dunes is often increased by residents through grading operations. Landscape projects using earth moving equipment are typically conducted from the houses outward, and sediment removed from the centers of the lots are usually placed near property lines. Many residents also prefer to push the sand which blows inland from the crest of the dune to the sides of their property, rather than push it back to the dune crest near the house where inundation will occur again. Public beach access ways are located along the lot lines; these rights-of-way are not graded by residents and they remain higher in elevation than the private property adjacent to them (Fig. 5). The natural vegetation (*Ammophila brevigulata*) extends landward from the foredune along the lot-lines. This grass, combined with exotic shrubs and the occasional sand fence, traps moving sand. The lot-line dunes, which occur where houses are located close to the beach, are major avenues for shore perpendicular aeolian transport. Burial of trees by wind blown sand at these locations provides evidence of rapid transport.

Figure 5 Lot-line dune at public beach access point between dwellings in Segment A.

At groins, both the dune and dune vegetation occupy a more seaward location than along the adjacent shoreline (Fig. 6). The dune may be higher on either the east or west side of the groin. Higher dunes on the east (updrift side, considering net transport) are attributed to the persistence of a wide beach, which provides a sediment source area and protects the existing dune from wave erosion. Higher dunes on the west side (Fig. 7) may be the result of the accumulation of wind blown sediment on the upwind side of either the groins or the dunes located at the groins. Pronounced temporary reversals of longshore transport may also occur (Panuzio 1968). These reversals result in the creation of wider beaches on the west side of the groins which temporarily protect incipient dunes and allow them to grow. Residents have also increased the height of dunes west (downdrift) of groins using sand fences to compensate for loss of protective beach. Dunes downdrift of groins

Figure 6 Segment A looking northwest, March 1980.

Figure 7 Eolian accretion on west (upwind) side of groin in Segment A. The west side is downdrift given the net direction of wave-induced longshore transport.

are usually steeper than updrift dunes because of truncation of the seaward portion of the dune by wave erosion (Fig. 6). Many of the dunes on the east side of groins in Segment A have profiles similar to the one portrayed in Figure 3 (profile A), whereas dunes on the west side resemble Figure 3 (profile B) or (profile C). The seaward portion of the dune in segment A is more crenulate in plan than in the other segments due to the salient formed by wave and wind accretion at the landward end of each groin.

Sediment starvation caused by the groins has accelerated erosion of the beach in Segment B. Erosion has proceeded to the point where the beach is too narrow to protect the dune from wave erosion by storms with a frequent recurrence interval. The high, narrow dune is obliterated, and subsequent storms occur too frequently to allow dunes to form by natural processes. Aeolian processes are only important up to this critical overwash event. Protective barriers to wave overwash may be created by earth moving equipment because of the speed at which sand may be emplaced. In Segment B, sand deposited landward of the beach on the main road along the island has been pushed seaward of the road to provide temporary protection against small storms. Through time, the active beach profile migrates to the location of the houses (Fig. 2) and even bulldozed dunes are no longer a viable option for shore protection.

Aeolian processes will again be important in creating a protective dune if sediment is replaced on the beach through artificial nourishment or if a natural beach profile can be re-established farther landward. The former alternative is being given serious consideration by the Corps of Engineers and local planners for Westhampton Beach, and it is likely that a beach nourishment project will be implemented in the future (Buckley 1980).

The effect of beach nourishment projects

Design criteria have not been developed for future beach nourishment projects, so it is not possible to calculate the magnitude of the effects. It is important to identify the potential effects, however, so that management guidelines may be specified prior to the operation. The width of the nourished beach and the characteristics of the fill sediments determine the value of the beach as a source of sediments for dune growth. Beach width also affects the frequency at which the dune will be attacked by waves. Studies indicate that there is a minimum source width below which aeolian transport will be less than the capacity for a given wind speed. This distance varies from 2 to 15 m in wind tunnels (Walker 1981); in the field, transport is believed to be fully developed at 5 m for dry sand and at 10 m for wet sand (3-4% water content) (Hotta 1984). Existing beaches exceed this width, so this variable alone would not have a significant effect on growth of the dune due to onshore winds. A wider beach will result in an absolute increase in the quantity of sand moved by winds blowing parallel or oblique to the shoreline, however, and there will be some additional accumulation upwind of groins, seaward of present deposits. Source width will thus be an important variable where obstructions to transport are in place perpendicular to the shore.

The primary significance of a nourished beach to the form of coastal dunes is the protection it provides against wave erosion. The buffer of beach sediments prevents existing dunes from being eroded and changes the location where fences are placed by the residents. As property owners place new rows of fences seaward of existing dunes, the dune will be transformed from a linear ridge to a wide platform. The surface of the dune platform will likely be variable in shape because owners will emplace fences and vegetation on an individual basis, not according to consistent practice. The form of the dune in nourished areas will again be similar to the dune in Figure 3 (profile A).

The sediment which will be used to nourish beaches at Westhampton Beach will come from adjacent inlets, bays, or offshore sources, where borrowed sediments will differ from native materials in grain size and sorting. Where coarse or heavy particles are included among finer or less dense particles, a surface lag may develop which will resist deflation. The armoring effect on the beach surface is enhanced if a large amount of shells comprise the beach materials (Carter 1977). Meteorologic conditions on Long Island favor a wave regime which results in short-term erosion-deposition cycles which will break up or bury the surface lag, increasing the likelihood of deflation and delivery of sediments to the dune. The effects of beach armoring by residual deposits is not significant under present conditions, but could be a factor if there is a sizeable coarse fraction in the fill material.

Sediment sorting is also significant in determining initial rates of transport from the nourished area. For example, using Bagnold's (1941) equation for eolian sediment transport, a 7.7 m/sec wind blowing across well sorted sand 0.4 mm average diameter will move sand at a rate of 18 911 cm^3/m of source area; if sediments with the same average diameter are poorly sorted, the calculated rate of transport is 35 301 cm^3/m, an increase of 86.7%. The rate of transport of poorly sorted sand with a diameter of 0.25 mm is only 18.2% greater than the rate for poorly sorted sand at 0.4 mm diameter. This indicates that the degree of sorting may be more important than grain size in determining rate of deflation of the fill material under certain conditions. Beach sands in the study area are well-sorted to moderately-well-sorted (Snyder & Pinet 1981). Eolian transport will increase if fill materials are more poorly sorted.

Enhanced eolian activity has been reported as a consequence of nourishment operations (Draga 1983). Sand fences should be placed at gaps in the dune line before beaches are nourished or the sediment will leave the project area. Eolian losses are rarely considered in determining the amount of fill required to create a beach of a given width and height. Eolian losses from a beach may be considerable, and they should be anticipated in advance. Sand fencing programs are often conducted on an ad-hoc basis and are not integrated with nourishment plans.

The principal considerations of beach nourishment to coastal dune formation are that nourishment provides a platform on which dunes can form, a new source of sediment to cause dunes to grow, and a buffer of sediment to protect the dunes from destructive storm wave attack. Under conditions of nourishment, dune changes on a developed barrier are cyclic, and these cycles are human induced. Attempts to short-circuit the cycle by creating dunes on narrow beaches with earth moving equipment will fail if beaches are too narrow to protect the new dunes from small storms.

Conclusions

This study is intended to complement existing studies of coastal dunes on developed barrier islands by providing quantitative estimates of the role of the wind in sediment transfers, a descriptive model of sequential landform development, and an assessment of the effects of shore protection alternatives. The results of the investigation are:

(a) The rate of sediment transport by waves exceeds the rate of transport by wind by several orders of magnitude.

(b) Wave erosion of the dune, followed by aeolian transport from the beach to the new beach/dune contact, is primarily responsible for the landward displacement of the dune line. Aeolian transport to the base of the dune exceeds transport over the crest by an order of magnitude.

(c) The value of dunes for human needs is for protection, and fences and dune grass plantings will be implemented to retain a dune between the ocean and the first row of houses.

(d) Sand fences are usually oriented parallel to the beach. Sand transport calculations show that maximum aeolian transport is alongshore, so sand trapping efficiency may be improved by re-orienting or re-configuring the fences.

(e) Foredunes seaward of houses acquire a characteristic "knife-edge" form, resulting from homeowner attempts to hold the dune crest at a specific position. These steep dunes are less stable than broader dunes.

(f) The presence of a house immediately behind a foredune alters wind and sand transport patterns. Scour occurs on the back of the dune adjacent to the dwelling, resulting in a crenulate back-dune form.

(g) Residential lot-lines are narrow zones of accumulation which favor aeolian transport landward of the dune crest line. Lot-lines dunes are primarily a result of landscape efforts on the inner lot, leaving a ridge and dune vegetation as a narrow buffer between dwellings.

(h) Groins reduce the rate of loss of sand from the beach where they are emplaced, and the dunes within the groin field pass through the sequential stages of evolution at a slower rate than in adjacent unprotected segments. The seaward line of the dune crest is more irregular in plan than in undeveloped areas. Dunes on the downdrift side of groins are steeper than on the updrift side.

(i) Aeolian activity plays a role in dune formation until storm wave erosion occurs too frequently to allow a dune to grow. Aeolian activity will again be significant if the beach is artificially nourished. A new cycle of changes will then occur.

(j) Specification of fill for beach nourishment projects has not included impacts of wind transport. Fine sediments will be rapidly moved upon emplacement, and fences and vegetation should be in place to prevent loss of fill. The significance of sediment sorting as well as that of grain size should be evaluated.

Acknowledgement

This research was supported by the National Park Service North Atlantic Region, Boston Massachusetts (Contract cx1600-2-008).

References

Baccus, J.T. and J.K. Horton 1980. Pedestrian impacts: Padre Island. In *Barrier island forum and workshop*, B.S. Mayo and L.B. Smith Jr. (eds.), 89-102. Boston, MA: U.S. National Park Service, North Atlantic Region.
Bagnold, R.A. 1941. *The physics of blown sand and desert dunes.* London: Methuen.
Buckley, P.A. 1980. Reconciling structural and non-structural solutions to coastal erosion problems. In *Barrier island forum and workshop*, B.S. Mayo and L.B. Smith, Jr. (eds.), 148-68. Boston, MA: U.S. National Park Service, North Atlantic Region.
Carter, R.W.G. 1977. The rate and pattern of sediment interchange between beach and dune. *Proceedings, Symposium on Coastal Sedimentology*, 3-34. Tallahassee, FL: Florida State University.
Coastal Engineering Research Center. 1977. *Shore protection manual.*

Vicksburg, MI: U.S. Army Corps of Engineers, CERC, Waterways Experiment Station.

DeWall, A.E. 1979. *Beach changes at Westhampton Beach, New York, 1962-73.* Fort Belvoir, VA: U.S. Army Coastal Engineering Center Misc. Report No. 79-5.

Draga, M. 1983. Eolian activity as a consequence of beach nourishment observations at Westerland (Sylt), German North Sea coast. *Zeitschrift fur Geomorphologie* Supplement-Band 45, 303-19.

Edelman, T. 1972. Dune erosion during storm conditions. *Proceedings 13th Conference on Coastal Engineering,* 1305-11. New York: American Society of Civil Engineers.

Gares, P.A. 1983. Beach/dune changes on natural and developed coasts. *Coastal Zone '83,* 1178-91. New York: American Society of Civil Engineers.

Gares, P.A., K.F. Nordstrom and N.P. Psuty 1980. Delineation and implementation of a dune management district. *Coastal Zone '80,* 1269-88. New York: American Society of Civil Engineers.

Godfrey, P.J., S.P. Leatherman and P.A. Buckley 1978. Impact of off-road vehicles on coastal ecosystems. *Coastal Zone '78,* 581-600. New York: American Society of Civil Engineers.

Heikoff, J.M. 1976. *Politics of shore erosion: Westhampton Beach.* Ann Arbor, MI: Ann Arbor Science Publishers, Inc.

Hotta, S. 1985. *Wind blown sand on beaches.* Unpublished Ph.D. Thesis,Department of Engineering, University of Tokyo.

Kadib, A.L. 1964. *Calculation procedure for sand transport by wind on natural beaches.* U.S. Army Coastal Engineering Center, Misc. Paper No. 2-64.

Knutson, P.L. 1978. Planting guidelines for dune creation and stabilization. *Coastal Zone '78,* 762-79. New York: American Society of Civil Engineers.

Leatherman, S.P. 1980. Pedestrian impacts: Assateague Island. In *Barrier island forum and workshop,* B.S. Mayo and L.B. Smith, Jr. (eds.), Boston, MA: U.S. National Park Service, North Atlantic Region.

Long Island Regional Planning Board. 1984. *Hurricane damage mitigation plan for the south shore of Nassau and Suffolk Counties, New York.* Hauppauge, NY.

McCluskey, J.M., K.F. Nordstrom and P.S. Rosen 1983. *Analysis of -eolian factors: Fire Island to Montauk Point, NY.* National Park Service/Rutgers University Cooperative Research Unit Technical Report.

McCormick, C.L. 1973. Probable causes of shoreline recession and advance on the south shore of eastern Long Island. In *Coastal geomorphology,* D.R. Coates (ed.), 61-71. Binghamton, NY: State University of New York Publications in Geomorphology.

Nordstrom, K.F. and J.M. McCluskey 1984. Considerations for the control of house construction in coastal dunes. *Coastal Zone Management Journal* 12, 385-402.

Nordstrom, K.F. and J.M. McCluskey 1985. The effects of houses and sand fences on the eolian sediment budget at Fire Island, New York. *Journal of Coastal Research* 1, 39-46.

Panuzio, F.L. 1968. The Atlantic coast of Long Island. *Proceedings, Eleventh Conference on Coastal Engineering*, 1222-41. New York: American Society of Civil Engineers.

Research Planning Institute. 1983. *Fire Island Inlet to Montauk Point, Long Island, New York, Reformulation Study, Sediment Budget Analysis Summary Report*. New York: U.S. Army Engineer District, NY.

Rosen, P.S. 1979a. Aeolian dynamics of a barrier island. In *Barrier islands*, S.P. Leatherman (ed.), 81-98. New York: Academic Press.

Rosen, P.S. 1979b. An efficient, low cost, aeolian sampling system. *Current Research*, Part A. Geologic Survey of Canada, Paper 78-1A.

Savage, R.P. and W.W. Woodhouse, Jr. 1968. Creation and stabilization of coastal barrier dunes. *Proceedings, Eleventh Conference on Coastal Engineering*, 671-700. New York: American Society of Civil Engineers.

Snyder, M.R. and P.R. Pinet 1981. Dune construction using two multiple sand-fence configurations: implications regarding protection of eastern Long Island's south shore. *Northeastern Geology* 3, 225-9.

Taney, N.E. 1961. *Littoral materials of the south shore of Long Island, New York*. Washington, D.C.: U.S. Army Corps of Engineerings, Beach Erosion Board, Technical Report 128.

Tsoar, H. 1984. Wind tunnel modelling of echo and climbing dunes. In *Eolian sediments and processes: developments in sedimentology* 38, M.E Brookfield and T.S. Ahlbrandt (eds.), New York: Elsevier.

United States Army Corps of Engineers. 1980. Beach erosion control and hurricane protection project, Fire Island Inlet to Montauk Point, 1:2,400 scale photogrammetric survey maps. U.S. Army Engineer District, NY.

Van de Graaf, J. 1977. Dune erosion during a storm surge. *Coastal Engineering* 1, 99-134.

Walker, J.D. 1981. *Eolian experimental study of wind ripples*. Unpublished M.Sc. Thesis, Department of Geology, Massachusetts Institute of Technology.

9

Quaternary dunes of the Pacific Coast of the Californias

Antony R. Orme and Vatche P. Tchakerian

Abstract

Along the Pacific coast of the Californias, recurrent dune deposition during Quaternary times has been favored by the localized presence of subsiding basins. Two such basins - the Santa Maria Basin in California and the San Quintín Basin in Baja California - are selected for detailed study. In terms of age, four dune phases are recognized: Phase I paleodunes of mid-Pleistocene age and older; Phase II paleodunes of late Pleistocene and early Holocene age; Phase III parabolic dunes that formed after 6000 ^{14}C years BP and became relict as sand supplies were curtailed after the main Flandrian transgression; and Phase IV dunes that expressed renewed aeolian activity after 1800 ^{14}C years BP. Similarities shared by each dune phase include orientation, internal structure, grain-size modality, grain shape, grain physical microfeatures, and detrital mineralogy. Significant differences, which permit discrimination between dune phases on the basis of age, include surficial morphology, grain-size distribution, grain chemical microfeatures, weathering mineralogy, and vegetation characteristics.

Introduction

Recurring phases of coastal dune development during Quaternary time have been recognized in several parts of the world, for example along the coasts of eastern Australia (Pye 1983, Thompson 1983), southeast Africa (Orme 1973, Hobday & Orme 1975), and north-central Chile (Paskoff 1970). This ignores the extensive dunes that often reach the coast from the world's major deserts, for example in Mauritania and Namibia, but which are not strictly

coastal in provenance. Geomorphic, pedologic and stratigraphic
evidence all indicate phases of coastal dune formation interspersed
with episodes of weathering and erosion. Several distinct dune
phases are often recognized and placed in a relative age sequence,
but the absolute age of individual phases is largely conjectural
although radiometric dates are sometimes available for Holocene
dunes (Thom *et al.* 1978).

Quaternary dunes of various ages occur at intervals along the
Pacific coast of the Californias between 28° and 38°N. The larger
dune fields occur in coastal lowlands; smaller accumulations are
found where protruding headlands obstruct the prevailing northwest
winds, but aeolian deposits are sparse along intervening
mountainous coasts. Recognizable surface dunes, as opposed to
aeolian sands embedded in stratigraphic sequences, range in age
from at least mid-Pleistocene to present. For this paper, four
dune phases are recognized, although the older two phases
probably embrace several aeolian episodes. The oldest paleodunes
(Phase I) are found much modified on emergent marine terraces or
in deep subsiding basins. Late Pleistocene and early Holocene
paleodunes (Phase II) mantle the larger coastal lowlands, descending
below present sea level at the coast. Late Holocene features
embrace Older Flandrian dunes (Phase III), formed during and after
the Flandrian transgression, and active Younger Flandrian dunes
(Phase IV) along the present coast. California's coastal dunes have
been described generally, notably by Cooper (1967), but those of
Baja California have not been closely examined, although Inman *et
al.* (1966) discussed the active dunes fronting the Vizcaino Basin.
There has been no serious attempt to explain the localized
persistence of these dunes, nor to define the similarities and
differences exhibited by successive dune phases.

This paper first explains the localized persistence of coastal
dunes, primarily relative to the availability of subsiding basins.
Quaternary coastal dunes occur in six principal localities along the
coast, of which two - the Santa Maria Basin in California and the
San Quintín Basin in Baja California - are selected for detailed
study. The similarities and differences between the four dune
phases are then examined. The paper's main objective is to provide
a temporal perspective to coastal aeolian geomorphology by
addressing those conditions that favor recurrent dune development,
and by distinguishing between those dune characteristics that are
inherited upon stabilization, and henceforth timeless, and those that
change with age and are thus time-dependent.

Persistence of dune development

The two main needs for coastal dune development, namely sand
supply and effective onshore winds, are met throughout the study
area and will be discussed only briefly. Why then is persistent

dune development so localized? The explanation, as will be shown, lies in the tectonic behavior of the coast during Quaternary time.

Sand supply and effective onshore winds
During Quaternary time, the coastal zone of the Californias received abundant clastic debris from the erosion of recently uplifted or still tectonically active mountains underlain by quartz-rich Mesozoic batholiths and later sedimentary formations. Most debris was flushed seawards onto the continental shelf and borderland where it was further distributed, generally southward, by waves and currents. Where shallow offshore areas were later exposed by tectonism and eustatic drawdown, these sediments were exposed to direct wind action or driven upslope by subsequent marine transgressions. Bathymetric and sampling data show such source areas extending upwind for 10-70 km northwest of the principal dunes. The largest source areas lie off San Francisco's Golden Gate and northwest of the Vizcaino Basin (Fig. 1). Elsewhere, despite the relatively narrow shelf north of Point Conception and the rugged borderland terrain farther south, restricted source areas were largely offset by the large volumes of clastic sediment flushed from coastal watersheds, especially during wetter episodes of the Quaternary. Sediment mineralogy will be discussed later.

Data from National Weather Service stations in California and local stations in Baja California emphasize the dominance of winds from the northwest quadrant, reflecting the predominant atmospheric circulation as modified by surface relief and thermal influences, the latter producing sea breezes during summer afternoons that often exceed the threshold velocity for sand transport. Although these northwest winds are approximately reflected by dune orientation, such data have several limitations. For example, data from Grover City, California, reveal that 50% of winds over 4.5 m s^{-1} blow from 280-300°, although nearby dune orientation is 310°. Thus, the effective wind direction, a mean value derived from the measured long axes of linear and parabolic dunes, may be a useful alternative reflection of the resultant sand transport vector responsible for dune movement. For dunes between Point Reyes and Point Conception, Cooper (1967) found that the effective wind varied from 300 to 321°, averaging 312°. For northern Baja California, excluding the Vizcaino Basin, the effective wind varies from 290 to 325°, averaging 304°.

Routine wind data from anemometer masts beyond dune fields also yield misleading indications of wind velocities at the surface of dunes. Mulligan (1985) has convincingly shown that wind shear velocities, and thus flow acceleration and sand transport rates, are much greater close to the dune surface than velocities predicted assuming a logarithmic wind profile below the 10 m anemometer mast at Grover City. Based on carefull field measurements and theoretical observations, he estimates that the annual loss of sand along a 7.5 km front from Pismo Beach to the Santa Maria dunes is about 300 000 m^3 yr^{-1}, much higher than Bowan and Inman's (1966) estimate of 38 000 m^3 yr^{-1}.

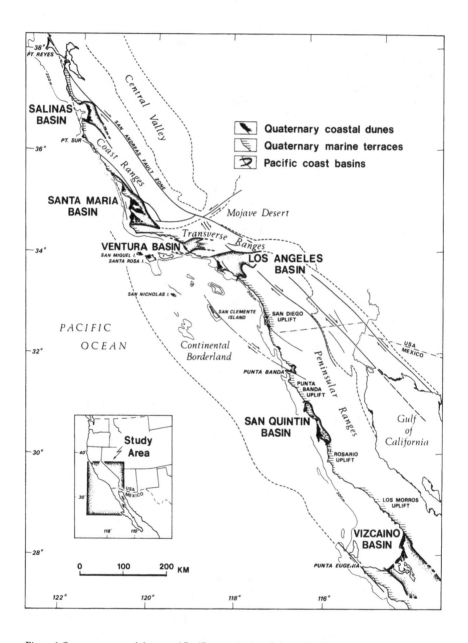

Figure 1 Quaternary coastal dunes and Pacific coast basins of the Californias.

Coastal tectonics and Quaternary dunes

To correlate tectonism and aeolian deposition may seem unusual, but the magnitude and persistence of Quaternary dune development along the Pacific coast appear to relate to patterns of tectonic behavior. In essence, recurring deposition has taken place where faulted or synformal basins have continued to provide subsiding areas near sea level. By contrast, in uplifted areas, aeolian deposition has been poorly developed and is expressed, if at all, by small ridges and thin cover sands on emergent marine terraces.

The coast of the Californias occurs where the Pacific plate emerges from the ocean and extends eastward to its margin along the San Andreas transform zone (Fig. 1). Coastal tectonism is closely related to this transform, although differently expressed in the miniplates that locally comprise the Pacific plate, namely the Coast Ranges, Transverse Ranges, and Peninsular Ranges and continental borderland. Widespread structural features within each miniplate reflect subordinate contractional or extensional components of motion along the transform (Dickinson 1981). Contractional effects include en echelon wrench folds and compressional basins, especially in the Coast Ranges. Extensional effects include the formation of pull-apart basins, especially within the Peninsular Ranges and continental borderland. Transient and local episodes of basinal subsidence also occur. Whereas these movements were Neogene in origin, their significance to this study is that basinal subsidence has continued, more or less, throughout Quaternary times.

The Coast Ranges contain two basins in which Quaternary aeolian deposition has been significant, namely the Salinas Basin and the Santa Maria Basin (Fig. 1). The NW-SE trending Salinas Basin is the result of compressional warping linked to wrench tectonics (Page 1981). Initially, the locus of marine deposition, strong crustal uplift over the past 4 million years, has caused this emerging depression to begin filling with synorogenic terrestrial sediments including coastal dunes. Strongly cross-bedded aeolian sands of probable mid-Pleistocene age occur up to 20 km inland. Continuing basinal subsidence is revealed by terrestrial sediments 900 m deep offshore (Gribi 1963), while marginal uplift is reflected in marine terraces flanking the Santa Cruz Mountains to the north which have been arched and progressively tilted seaward at a rate of 0.16-0.26 mm yr^{-1} (Bradley & Griggs 1976).

In contrast, the Santa Maria Basin *sensu lato* is a Neogene pull-apart structure between the southernmost Coast Ranges and the western Transverse Ranges, bounded by active strike-slip faults extending northwest to the Hosgri fault zone offshore. Separation began about 12 million years ago, and the basin became the locus of Neogene marine deposition (Hall 1978). Quaternary crustal shortening then raised the area above sea level, but continuing basinal subsidence has maintained conditions suitable for recurring coastal dune development.

The Ventura and Los Angeles basins are both ancestral parts of the continental borderland trapped against the flanks of the emerging Transverse Ranges (Howell & Vedder 1981). The narrow Ventura Basin results from massive Neogene compression within the rotating Transverse Ranges which has produced an elongate fault-bounded synformal corridor extending 200 km west from the San Gabriel Fault to Point Conception. Basinal subsidence and marginal uplift were more intense here than in other basins. Fission-track, radiometric and paleomagnetic age determinations in marine sedimentary rocks reveal that the Ventura Basin subsided at rates of 9.5 ± 2.5 mm yr^{-1} from 2.0 to 0.6 million years ago, since which time the basin's margins have risen at a mean rate of 10 mm yr^{-1} (Yeats 1978). Over 7000 m of marine and non-marine, including aeolian, sediments of Plio-Pleistocene age underlie the Oxnard Plain, aeolian sands being of small extent at the coast.

The Los Angeles Basin may have formed in a pull-apart system between the continental borderland and the San Gabriel-San Andreas transform. Quaternary subsidence accompanied compression and slivering as the Peninsular Ranges underthrust the Transverse Ranges along the basin's northern margin. Within the basin, coastal aeolian sands are interbedded with alluvial deposits, but the later Quaternary dunes developed along a front 30 km long and 10 km deep behind Santa Monica Bay have been largely obliterated by urban and industrial development.

In Baja California, extensive Quaternary dunes occur in three basins which also originated as ancestral parts of the continental borderland, namely the San Quintín, Vizcaino and, beyond our study area, Magdalena basins. The San Quintín Basin, which extends about 100 km along the western flanks of the Peninsular Ranges, was first recognized during investigations of Quaternary marine terraces (Orme 1980). The terrace sequence revealed progressive coastal downwarping between flanking uplifts, the highest terrace reaching 345 m on Punta Banda to the north, and 357 m near El Rosario to the south, but descending below sea level north of San Quintín (Fig. 1). Continuing deformation and faulting, as well as Holocene volcanism, occur within this basin.

The Vizcaino Basin, 80 km wide and 200 km long onshore, lies between the peninsular's spine and Sierra Vizcaino. Along its eastern margin, marine terraces indicate Quaternary uplift increasing northward (Woods 1978, Orme 1980). Sierra Vizcaino is flanked by similar terraces rising to 250 m which suggest an uplift rate of 0.07 mm yr^{-1} (Ortlieb 1978). The 5-7 m terrace, about 120 000 years old, is relatively undeformed around this basin, suggesting that marginal uplift ceased some time ago although some basinal subsidence may have continued. The basin contains interesting late Quaternary aeolian, lagoonal and alluvial deposits, merging southeast into the Vizcaino Desert proper. Studies of the most recent Holocene features have shown that the present lagoons began forming some 6000-7000 years BP while the present barrier beach, on which active dunes up to 3 km wide occur, formed about 1800 years BP (Phleger & Ewing 1962, Inman *et al.* 1966). Coastal dunes also occur northward on low protruding headlands at Punta Rosallilita, Punta Rocosa, Punta Negra and Punta Maria.

The offshore zone is of interest, partly because it provided the immediate sand supply for coastal dunes, and partly because some islands experienced recurrent aeolian deposition with subsequent formation of paleosols and calcareous eolianites. The fault-bounded ridges and basins of the continental borderland developed in a Neogene wrench-tectonic setting at the evolving margin between the Pacific and American plates. As ridges rose above a particular Quaternary sea level, they were draped in aeolian sand rich in shell fragments and other calcareous matter. San Miguel, Santa Rosa, San Nicolas, Santa Barbara and San Clemente islands all carry eolianites or paleodunes in various stages of weathering and dissection, whereas Santa Cruz and Santa Catalina islands largely escaped such deposition, possibly because of their relatively late emergence. Age determinations on marine terrace assemblages on some of these islands indicate different uplift rates and rate changes from island to island over the past 0.5 million years. In this disharmonic tectonic setting, the presence or absence of aeolian deposits may provide a key to an island's Quaternary history.

The Santa Maria and San Quintín dune complexes

To evaluate similarities and differences between successive dune phases, the aeolian complexes of the Santa Maria and San Quintín basins are investigated. Relative and absolute dating techniques permit some age characterization of the four dune phases recognized.

The Santa Maria Basin contains three dune complexes and four dune phases ranging on stratigraphic evidence from at least mid-Pleistocene time to the present. The Morro complex extends 14 km alongshore and 8 km inland; the Santa Maria complex, 28 km alongshore and 50 km inland; and the Santa Ynez complex, 30 km alongshore and 42 km inland (Fig. 2). Each complex covers a triangular area in a fault-controlled basin from which dunes have been partly removed by subsequent fluvial erosion. Phase I paleodunes occur within the Orcutt Sand, a complex of aeolian and shallow-water sediment which overlies and interfingers with earlier Pleistocene fluvial deposits. Aeolian components within the upper Orcutt Sand are poorly consolidated, oxidized sands with iron pisoliths and much clay. These sands may once have blanketed the Santa Maria Basin, but have been partly eroded following uplift of the Casmalia Hills and San Luis Range. The Orcutt San is subject to ongoing investigations not reported here. Phase II paleodunes, stratigraphically late Pleistocene to early Holocene in age, partly

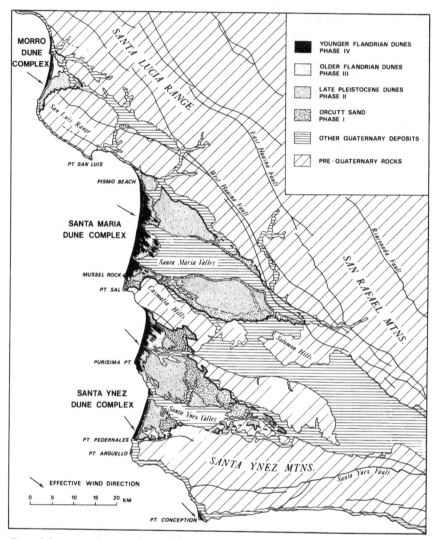

Figure 2 Quaternary dune complexes of the Santa Maria Basin, California.

mantle the Orcutt Sand, but in turn disappear beneath Flandrian sands toward the coast. These dunes cover the Nipomo and Orcutt mesas north and south of the Santa Maria valley respectively, and are composed of poorly consolidated to unconsolidated, red to yellow sands. Older Flandrian Phase III dunes form impressive parabolic dunes between Pismo Beach and the Santa Maria River. Younger Flandrian Phase IV dunes are active transverse forms extending up to 3 km inland from the coast, though active sands and reactivated Phase III materials occur up to 5 km inland (Fig. 3).

Figure 3 The Santa Maria dunes extending north for 20 km from the Santa Maria River, showing active Phase IV dunes, parabolic Phase III dunes immediately to the east, and Phase II paleodunes of wooded Nipomo Mesa to the northeast, November 1947 (Spence Collection, UCLA).

At least four dune phases are also found in the San Quintín Basin (Fig. 4). Phase I paleodunes are visible dune remnants found 100-300 m above sea level on higher marine terraces extending up to 10-15 km inland from the present coast. On paleontological grounds, the terraces are of mid-Pleistocene age and older (Hertlein & Allison 1959, revised), and the overlying, indurated dune sands have survived owing to the increased aridity of later Pleistocene time. Phase II paleodunes are found on a marine terrace which descends from 4 m above sea level at El Consuelo to pass gently beneath the alluvial Valle San Quintín farther north. This terrace has been dated around 120 000 years BP (Orme 1980), and some of its aeolian sands may date from this time. However, the downwarping and eventual disappearance of this terrace beneath Valle San Quintín suggests that the extensive Phase II paleodunes there may range from late Pleistocene to early Holocene time. Along the eastern shores of the San Quintín lagoon, Pismo clams (*Tivela stultorum*) from midden deposits immediately overlying the reddish sands of these paleodunes have been dated by ^{14}C at 6165 ± 250 and 6055 ± 250 years BP (Hubbs *et al.* 1962). Similarly, a rock scallop (*Hinnites multirugosus*) from a midden deposit overlying paleodune sand at Punta Cabras, 100 km NNW, yielded a ^{14}C date of 6400 ± 200 years BP (Hubbs *et al.* 1960). These dates place an effective upper limit to the likely age of the Phase II paleodunes,

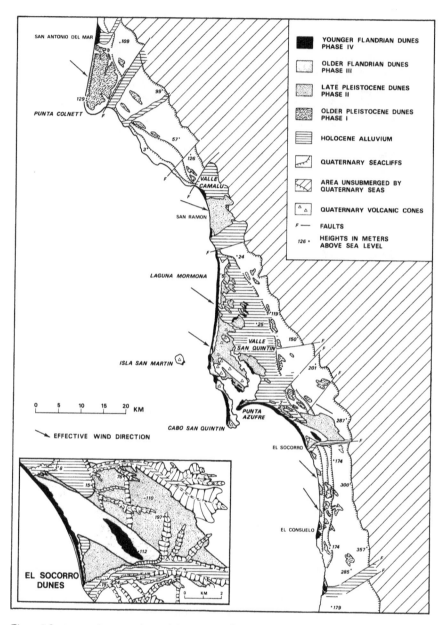

Figure 4 Quaternary dune complexes of the San Quintín Basin, Baja California, Mexico.

at about the time the Flandrian transgression was rising close to present sea level, and the early inhabitants were establishing fishing camps along the submerging shore. Nevertheless, this is but an upper limit, and it is very possible that the dunes are much older and that fishing folk were simply camping on relatively resistant coastal bluffs composed of paleodune material. More significantly, these dates place a lower age limit on the Older Flandrian Phase III dunes which overlie the Phase II dunes and the midden deposits, but which descend below sea level at El Socorro and elsewhere in the region. These Phase III dunes, which extend for 25 km north from Cabo San Quintín behind the modern dunes, were probably initiated during the Flandrian transgression, and as sand supplies were progressively curtailed by rising sea level, became stabilized by vegetation. Younger Flandrian Phase IV dunes are mostly narrow active ridges that have formed since the close of the main Flandrian transgression. Pismo clam valves and charcoal from a midden clambake immediately below grey Phase IV sand at San Antonio del Mar have yielded ages of 1800 ± 300 and 1660 ± 200 years BP respectively (Hubbs *et al.* 1965). It may be remembered that the barrier beach at the north end of the Vizcaino Basin also began forming around 1800 years BP. In the San Quintín Basin, therefore, there is relatively good control over the dating of the later dune phases, and on morphological, textural and pedological grounds, it seems reasonable to extend these controls to the Santa Maria Basin. At San Quintín, volcanic flows and ejecta overlie Phase II paleodunes, but underlie Phase IV dunes. The more northerly cinder cones of the group are thought to be less than 3000 years old (Gorsline & Stewart 1962).

Similarities between dune phases

Similarities between successive dune phases are expressions of the original deposit that experience little significant change over time, and in that respect become timeless upon dune stabilization. Dune orientation, internal structure, dominant mineralogy, and those sediment properties indicative of the formative conditions fall into this category.

Dune orientation and internal structure
The surface orientation of the Younger Flandrian Phase IV dunes, reflecting effective winds between 300° and 321° at Santa Maria and between 290° and 325° at San Quintín, is also found in the Older Flandrian Phase III dunes and obscurely in the Phase II and Phase I paleodunes (Fig. 5). This observation has paleoclimatic significance. Whereas Quaternary climatic oscillations are known to have caused some latitudinal shift of certain temperature and precipitation parameters along the Pacific coast, it appears that effective dune-forming winds remained essentially unidirectional within the northwest octant. Internal structures such as strong cross-beds, indicative of aeolian transport, are also found in all dune phases, although much obscured by weathering in the paleodunes. Indeed, the presence of medium to large scale

Figure 5 Quasi-parabolic Phase III dunes emerging from beneath active Phase IV dunes on the barrier beach at Morro Bay, California, November 1947 (Spence Collection, UCLA).

cross-bed sets facing downwind, with foreset dips near angle of repose and sharp low-angle confining surfaces, is an important diagnostic feature which flags old sand deposits for further investigation as possible dunes in the absence of other evidence.

Grain size

Recognizing the visual differences between dune phases in the Santa Maria and San Quintín complexes, a stratified sampling plan was used to acquire 240 sediment samples, each weighing 60-80 g. For Santa Maria, 50 samples were then randomly selected from windward slopes of the Phase IV transverse dunes, 48 samples from the windward slopes of Phase III parabolic dunes, and 47 samples from the Phase II paleodunes. A smaller sample population was used to characterize the four phases at San Quintín. These samples were then sieved at half-phi (Santa Maria) and whole-phi (San Quintín) intervals, plotted, and statistical parameters calculated following the methods of Folk and Ward (1957). The data are summarized in Figures 6 and 7, and Table 1. The graphic representations show some expected similarities between dune phases, notably strong modality in the 2-3 phi class, and a mesokurtic to leptokurtic trend, but there are significant differences which will be explained later.

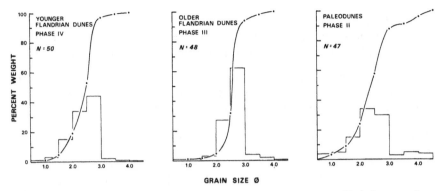

Figure 6 Grain-size histograms and cumulative curves, Phases II, III and IV, Santa Maria dune complex, California.

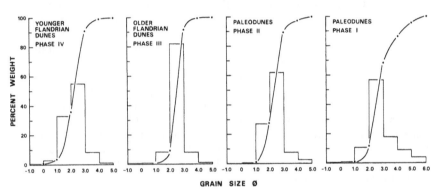

Figure 7 Grain-size histograms and cumulative curves, Phases I-IV, San Quintín dune complex, Baja California.

Table 1 Grain-size parameters, Santa Maria and San Quintín dune complexes.

Grain-size parameter	Dune complex	Dune phases			
		IV	III	II	I
Mean (M_z)ϕ	Santa Maria	2.28	2.46	2.26	
	San Quintín	2.20	2.50	2.28	2.88
Sorting (Σ_I)	Santa Maria	0.50	0.42	0.82	
	San Quintín	0.65	0.49	0.70	0.95
Skewness (Sk_I)	Santa Maria	−0.19	−0.14	0.04	
	San Quintín	−0.13	−0.08	−0.08	0.27
Kurtosis (K_G)	Santa Maria	0.92	1.10	1.32	
	San Quintín	0.94	1.24	1.17	1.18
% Silt	Santa Maria	0.1	1.0	5.9	
	San Quintín	0.7	0.6	2.4	13.9

Grain shape

For roundness and sphericity analysis, 100 quartz grains held by each 0.5 phi screen in the 1-4 phi range from Santa Maria were sprinkled on glass slides and examined under a binocular microscope. For roundness and sphericity indexing, the methods proposed respectively by Powers (1953) and Rittenhouse (1943) were used. Values, multiplied by 100, are shown in Figure 8.

Taking Phases II, III and IV together, the quartz grains were found to be mostly subangular to angular, with subrounded to rounded grains comprising only 11-18% of Phase IV sands, 12-17% of Phase III sands, and 18-24% of Phase II sands. Thus, unlike desert sands which are generally subrounded to well rounded from recurrent abrasion over long periods (Krinsley & Doornkamp 1973), these coastal grains indicate little rounding prior to deposition which, in turn, may reflect the limited abrasional time available in previous, rapidly changing nearshore and fluvial environments along a fast eroding continental margin. Contrary to the views of Shepard and Young (1961), roundness alone provides no adequate basis for distinguishing between local beach and dune sands.

Both roundness and sphericity do, however, decrease with diminishing grain size (Fig. 8). Larger grains are generally more readily abraded than smaller particles, as observed by Goudie and Watson (1981) and Jawad Ali and Ali-Ani (1983). Furthermore, similarity in grain shape relative to grain size holds true for all three dune phases, indicating that post-depositional events have little impact on general grain shape once sands have stabilized within a sedimentary mass.

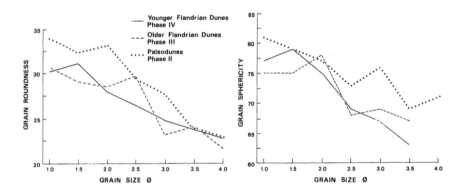

Figure 8 Grain size related to grain roundness and sphericity, Santa Maria dune complex, California.

Detrital mineralogy

Throughout the Californias, the sand fraction of nearshore sediments ultimately available for coastal dune development consists largely of detrital grains of quartz and feldspar, with some mica. The most abundant heavy minerals are magnetite and ilmenite, derived from the erosion of ferromagnesian rocks inland. Less abundant, but often associated with these black sands, are inert oxides and silicates such as chromite, rutile, zircon, monazite, uranoan thorite and garnet, together with some gold and platinum (Wilson & Mero 1966). As offshore depths increase, benthonic foraminifera tests and minor quantities of diatoms and radiolarians occur, while authigenic glauconite and phosphorite are found at and beyond the shelf edge. Shell fragments are ubiquitous.

The mineralogy of the Santa Maria dunes was analyzed by binocular microscope, using a random sample of 100 grains from each dune phase. Percent composition is shown in Table 2. Composition of the San Quintín dunes was very similar, with quartz ranging from 75 to 90%, orthoclase and albite important, and muscovite and shell fragments completing the light fraction. Of the heavier minerals, biotite, amphibole, pyroxene, sphene, epidote, garnet, zircon, ilmenite and magnetite were found. For both dune complexes, this composition is consistent with the source areas, but as will be shown, weathering has produced some significant changes.

Table 2 Mineral percentage compostion, Santa Maria dune complex.

	Dune phases		
	IV	III	II
Quartz	82	81	84
Feldspars*	13	14	9
Heavy minerals**	5	5	7

*   Of the feldspars, albite constitutes 70%, pink orthoclase 30%
**   Of the heavy minerals, magnetite, hematite, biotite, garnet, and amphibole are most common

Differences between dune phases

Despite similarities, each dune phase is distinguishable in terms of surface morphology, grain-size distribution, grain micromorphology, weathering mineralogy, and vegetation characteristics. Surface morphology reflects the intensity and duration of post-depositional weathering and erosion. Grain-size distributions are temporal responses to physical and chemical weathering, as also reflected in grain micromorphology and mineralogy. Vegetation changes are linked to plant successions that also reflect changing soil and shelter properties. Such differences provide important criteria for distinguishing dune phases over wide areas.

Surface morphology
Each dune phase is distinguishable in terms of mean slope, relative relief and dissection ratios, although weathering and erosion intensity vary along the coast.

Phase I dunes are moderately to well dissected sand sheets with low hummocks rarely over 5 m in relative relief. The hummocky Orcutt Sand rises to over 300 m above sea level in the Santa Maria Basin and is dissected by canyons up to 100 m deep. In Baja California, dune dissection on the higher terraces was accomplished by fluvial processes now rare to the area. Dissection is less advanced at lower levels, and at Punta Colnett a 3-20 m thick dune sheet covers 36 km^2 of marine terraces.

Phase II dunes are less dissected, occurring either as hummocky sheets with 5-20 m of relative relief or as isolated mounds over 20 m high. Hummocky sheets with vague parabolic or elongate forms cover 72 km^2 of Nipomo Mesa in the Santa Maria Basin and 35 km^2 at San Ramón, San Quintín Basin. Because they descend beneath present sea level, the maximum thickness of these dunes is uncertain, but dunes 75 m thick occur at San Ramón.

Flandrian dunes are either relict (Phase III) or active (Phase IV) forms little changed by non-aeolian forces. At Santa Maria, Phase III sands form parabolic dunes 300-1500 m long, 150-700 m wide, and 9-30 m high. Windward slopes are 12-25°, slip faces 29-33°. Crests are higher and rounder than in Phase IV dunes. These parabolic dunes are elongate U-shaped ridges, opened to windward by deflation which has scoured intervening troughs to within 2-3 m of mean sea level, thereby providing for shallow freshwater lakes and marshes (Fig. 9). At El Socorro, Phase III dunes form parallel ridges with intervening deflation corridors up to 30 m deep with a sand tongue, 8 km long and 2 km wide, which has invaded the nearby arroyo in a largely relict slip face, 32-35° steep and 60 m high. These dunes formed before the sea reached its present level, and were supplied by sand from the open coast across what is now a lagoon and bay (Fig. 3). Elsewhere, Phase III dunes emerge from beneath active Phase IV dunes as small ridges and parabolas, largely stabilized by vegetation, but locally reactivated by grazing and other disturbance (Fig. 5).

At Santa Maria, Phase IV dunes are active transverse forms aligned in three waves parallel with the shore and perpendicular to the effective 300° wind (Fig. 9). Windward slopes are 15-22°, slip faces 27-33°, and landward migration rates average 4.2 cm day^{-1} (Mulligan 1985). The three transverse waves are separated by part-vegetated depressions and further distinguished by morphology. Crest-to-crest distances are 5-15 m in the outer dunes nearest the coast, 10-35 m in the middle dunes, and 30-200 m in the inner dunes. Slip-face heights are 1-10 m, 1-14 m, and 4-22 m for the outer, middle and inner dunes respectively. Transverse dunes also cover an inland area 2 km long, 0.6 km wide, and up to 30 m high at El Socorro. At nearby Punta Azufre, transverse and barchanoid

Figure 9 Three waves of transverse Phase IV dunes overlapping parabolic Phase III dunes, with Phase II paleodunes in the distance, Santa Maria dune complex, California. (Photo: K.R. Mulligan)

forms 2 m high are migrating over a 5 km distance along the backshore. Elsewhere, Phase IV dunes often form narrow ridges on Flandrian barrier beaches. Extending 40 km north from Cabo San Quintín, for example, active dunes are 50-300 m wide and 9-12 m high, with cross-ridges and deflation corridors up to 10 m deep averaging 20 units per kilometer.

Grain-size distribution
Modal class apart, grain-size distributions of the four dune phases reveal important differences (Table 1, Figs 6 & 7). Phase I sands are relatively fine, moderately sorted, fine skewed, and leptokurtic. Phase II sands are slightly less fine, moderately sorted, near symmetric, and leptokurtic. Phase III sands are coarser than phases II or IV, well sorted, negatively skewed towards coarser grains, and leptokurtic to mesokurtic. Phase IV sands are coarser

than Phase III, moderately well sorted, negatively skewed, and mesokurtic.

Statistical analyses were conducted to evaluate further the significance of these observed differences for dune phases II, III and IV at Santa Maria. A one-way analysis of variance revealed, in order of importance, significant differences in sorting, skewness and kurtosis for all three phases.   Mean grain size differed significantly between Phase III dunes and the others.

Bivariate analysis revealed a significant distinction between Phase II and Flandrian dunes.   Paleodune sands are less well sorted, near symmetric or slightly skewed toward finer grains, and leptokurtic - essentially because they have more fines than Flandrian deposits. Paleodunes also have a coarser fraction, however, which maintains a mean grain size similar to Phase IV dunes.   Considering samples from all three phases together revealed no significant relationship between mean grain size and sorting, or between mean grain size and skewness.

Multivariate discriminant analysis was performed to see whether or not the three dune phases could be differentiated more accurately using the four textural parameters simultaneously. Analysis yielded the following two discriminant functions (canonical variables 1 and 2 in Figure 10):

$$DF_1 = (-7.33854\ \Sigma_I) + (-2.92217\ Sk_I) + (-9.35344\ K_G)$$

$$DF_2 = (+6.75654\ \Sigma_I) + (+4.06057\ Sk_I) + (-19.95972\ K_G)$$

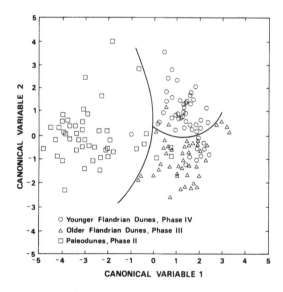

Figure 10 Scatterplot of discriminant analysis functions showing grain-size separation of Phases II, III and IV, Santa Maria dune complex, California.

This analysis revealed sorting to be the variable most responsible for differences between the three dune phases, followed by kurtosis and skewness. Mean grain size did not enter the analysis owing to its low tolerance level. Figure 10 shows that the three phases are clearly distinguished by considering textural parameters simultaneously, thereby comfirming discriminant analysis as a powerful tool for distinguishing between coastal dunes of differing ages. Greenwood (1969) and Moiola et al. (1974) have previously shown discriminant analysis to be an effective technique for distinguishing between aeolian and other sand bodies, while Moiola and Spencer (1979) have used the approach to differentiate between inland and coastal dunes.

Grain micromorphology
Exploring observed sediment differences further, quartz grains from phases II, III and IV at Santa Maria were examined under a scanning electron microscope using procedures described by Krinsley and Doornkamp (1973). Time-related changes in grain appearance from late Pleistocene to present time were noted and selected mechanical and chemical microfeatures for each phase are shown in Figure 11. Of the mechanical features, meandering ridges (Fig. 11A) are rare on Phase IV grains and not found on older grains. Such ridges, probably abrasion-related cleavages, are common to desert sands. Their scarcity here is probably due to the limited abrasional exposure of coastal dune sands compared to desert sands, a reason previously invoked to explain the paucity of rounded grains. V-shaped impact pits are rare to all phases, and because they are common to beach sands, may have been inherited from that source and survived. Conchoidal fractures (Fig. 11B), upturned plates, breakage blocks and jagged grain ends, all evidence of wind abrasion and percussion, are common to abundant in all phases, indicating little post-depositional modification. Adhered particles and clay platelets (Fig. 11C) are also common, but become more abundant in older dunes, suggesting grain translocation over time. Of the chemical features, solution pits and grooves are found in all phases, but most abundantly in paleodune grains (Figs 11D & E). Silica precipitation is rare on Flandrian grains, but abundant on paleodune sands. Most mechanical features are thus expressions of the original aeolian environment and have been modified little over time. Adhered particles, clay platelets, and various chemical features, which are more abundant in older dunes, are largely functions of time.

Weathering mineralogy
The strongly modal, well sorted Phase III sands indicate prolonged winnowing, but little weathering, and thus provide a basis against which to evaluate other phases. In physical terms, the coarser

grains of Phase IV sands indicate an active aeolian environment in which larger grains derived by traction from nearby beaches coexist with coarse lag deposits. Some of the adhered particles and clay platelets found in all phases, but most abundantly in paleodunes, may be introduced as airborne dust particles, a process that is cumulative over time. Dust settling on dunes adheres to grains in the presence of moisture and is washed downward by infiltrating rainwater. Muhs (1982) found silty aeolian dust driven by Santa Ana winds from the Mojave Desert to occur in soils on San Clemente Island, 200 km to the south. Similar dust influxes have probably occurred throughout the study area, partly explaining the fine tail of older dune sands.

In other respects, differences in grain-size distribution and micromorphology invite explanation in terms of post-depositional weathering. Chemical weathering may both increase and decrease grain size. Increased grain size results from the precipitation of silica, iron, and carbonate solutions onto grains. Silica precipitation, abundant on paleodune grains (Fig. 11F), is presumably effected by vadose water which dissolves quartz (see Figs 11D & E) and silicate minerals, and then promotes reprecipitation during evaporative periods (Al-Saleh & Khalaf 1982). Solution also liberates iron from various ferromagnesian minerals such as biotite, hornblende, epidote, pyroxene, garnet, ilmenite and magnetite, which are common to the area's heavy minerals and susceptible to hydrolysis. Subsequent precipitation provides the characteristic iron-oxide coatings found on grains, and thereby the yellow, brown and red color so typical of paleodunes. Under favorably quiescent conditions, iron oxides precipitate concentrically on sand grains to form iron pisoliths which may grow to over 10 mm in diameter. Iron pisoliths are particularly common in the uppermost 1 m of red aeolian and nearshore sands found on Pleistocene marine terraces near San Diego and on Punta Colnett. Red coastal sands have attracted much attention, as recently reviewed by Gardner and Pye (1981), but their paleoclimatic significance remains enigmatic. Certainly, coloration alone is no adequate indicator of age for much depends on initial sand mineralogy, aeolian dust inputs, temperature, and available moisture, the latter within the study area reflecting annual precipitation which ranges from 800 mm in the north to less than 100 mm in the south. Indeed, as aridity increases, iron oxidation is less notable, and carbonate precipitation occurs on grains as a prelude to various stages of calcrete formation.

Decreased grain size is largely produced by mineral disaggregation which particularly affects feldspar, with both orthoclase and albite releasing clay minerals and clay-size particles to the tail of the distribution, a process again influenced by available moisture and temperature conditions. Aeolian dust may also introduce silts and clays to the fine fraction, as suggested by Muhs (1982) for San Clemente Island. The significant tail of fines observed in the paleodunes, which becomes more pronounced with increasing age (Fig. 7), is clearly dependent in part on time, but so many other variables are involved, including the added problem of climatic change, that little more can be said at the present state of knowledge.

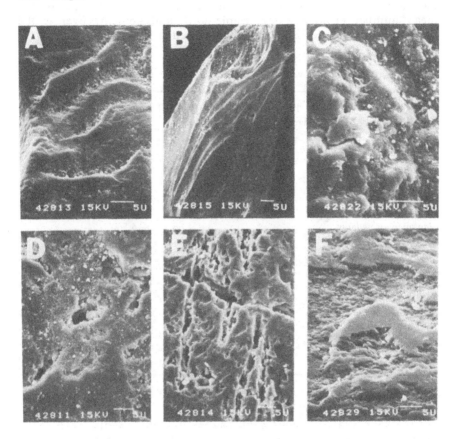

Figure 11 Scanning electron micrographs of selected quartz grain microfeatures, Santa Maria dune complex, California, showing (A) meandering ridges, Phase IV, (B) conchoidal fractures, Phase IV, (C) adhered particles and clay platelets, Phase III, (D) solution pits, Phase II, (E) solution grooves, Phase II, and (F) silica precipitation, Phase II.

Vegetation characteristics

Apart from surface morphology, perhaps the most visual difference between dune phases lies in vegetation characteristics which, though varying along the coast, are related to the complexities of plant succession and soil development. To test this observation, the three dune phases close to one another at El Socorro were chosen for detailed vegetation analysis (Fig. 4 inset). Fifteen individual plants were selected for study based on their dominance in one or more dune phase. Four cactus genera were subsequently grouped under Cactaceae, while grasses were recorded but not differentiated. Plant data were acquired by stratified random-walk, sampling with a 12.6 m^2 hoop 25 times in each dune phase. The data are summarized in Table 3, in which plant density records the mean number of plants per sample unit, and percent cover is the mean percent of the sampled area covered by the aerial parts of a plant.

Table 3 Plant density and cover, El Socorro dune complex, Baja California.

Taxon	Phase IV dunes Density	% Cover	Phase III dunes Density	% Cover	Phase II dunes Density	% Cover
Abronia maritima	1.96	3.13				
Lycium brevipes	0.92	1.36	0.28	0.16	0.96	5.12
Ephedra californica	0.12	0.01	0.16	0.37	0.04	0.04
Atriplex julacea			1.48	2.10		
Helianthus niveus			4.44	1.64		
Haplopappus venetus			4.68	4.56		
Cactaceae			0.08	0.01	3.44	3.65
Simmondsia chinensis			P	P	1.80	2.08
Euphorbia misera			P	P	0.76	3.05
Franseria chenopodifolia			P	P	1.56	4.10
Dudleya spp.					1.28	0.39
Agave shawii					1.16	3.03
Gramineae					13.88	3.68
Other spp.	0.08	0.03	4.04	3.69	2.32	4.28
Total	3.08	4.53	15.16	12.53	27.20	29.42

P = present, but cover and density less than 0.01, included with other spp.

Table 3 reveals a significant change in both plant density and plant cover from active Phase IV dunes, through Phase III, to the Phase II paleodunes, accompanied by a replacement of earlier colonists by invaders from the coastal scrub. These features are consistent with the facilitation mechanism of plant succession trending toward quasi-equilibrium with the regional environment. Localization of the foredune pioneer *Abronia maritima* within the Phase IV dunes is emphasized, together with its dominance in that habitat. Other Phase IV plants, primarily *Lycium brevipes* and *Ephedra californica*, also occur in the older dunes, but the latter is always localized in moister swales between sandy ridges. *Lycium brevipes* is among the most tolerant coastal shrubs, occupying all habitats except the youngest foredunes.

At El Socorro, *Atriplex julacea*, *Helianthus niveus* and *Haplopappus venetus* are confined to Phase III dunes, but there is little reason why they should not colonize Phase IV dunes. *Haplopappus venetus* is particularly dominant in Phase III dunes because grazing over the past few decades has created unstable secondary habitats for which it is well suited. Most other species found in these dunes are representatives of the coastal scrub.

Plants of Phase II paleodunes are reasonably typical of the local coastal scrub beyond the dunes, where *Lycium brevipes*, *Euphorbia misera*, *Franseria chenopodifolia* and *Agave shawii* compete for dominance. Grasses are also significant while the Cactaceae are a measure of past human and animal disturbance.

Plant densities increase from 3.08 in Phase IV dunes, through 15.16 in Phase III dunes, to 27.20 in Phase II paleodunes - a value not far removed from that of the regional coastal scrub beyond the dunes. Mean percent cover increases from 4.53 in Phase IV, through 12.53 in Phase III, to 29.42 in Phase II. These values at El Socorro are reasonably typical of the San Quintín dune complex, and provide a useful example of plant differences on dunes of different ages.

Conclusions

Along the Pacific coast of the Californias, between 28° and 38°N, recurrent dune deposition during Quaternary times has been favored by the localized presence of subsiding tectonic basins, while aeolian sedimentation has been limited in intervening uplifted areas. Quaternary coastal dunes of the Santa Maria and San Quintín basins provide information of wider application throughout this coastal zone. Four dune phases are recognized at the surface, for which relative and absolute dating techniques permit some age characterization.

Phase I dunes are strongly weathered and variably indurated sands of mid-Pleistocene age and older which occur on emergent marine terraces or in deep subsiding basins. Phase II dunes overlie a marine terrace dated around 120 000 years BP and also mantle larger coastal lowlands, descending below present sea level at the coast. Midden deposits about 6000 ^{14}C years old cap these dunes, but while some sands may be as young as early Holocene, a late Pleistocene age seems more appropriate for the bulk of these deposits.

Phase III dunes, which formed after 6000 years BP, and thus towards or after the close of the Flandrian transgression, often form well-preserved parabolas and linear ridges. At the changing unstable shoreface of the Flandrian transgression, with abundant source material moving onshore, dunes were well developed. As sand supplies were curtailed after the close of the main transgression, dune development became progressively restricted, some erosion occurred, but vegetation exerted a stabilizing influence. Some time after 1800 ^{14}C years BP, renewed dune development occurred and Phase IV dunes began forming the relatively narrow active ridges and larger transverse forms found along the present coast. The reason for this renewed aeolian activity is unclear, but it may be related to renewed shoreface erosion caused by a further rise of sea level attributable to glacioeustatic, hydroeustatic, or tectonic forces. The present barrier beach across the Vizcaino Basin also began forming around 1800 years BP. This Holocene dune activity resembles that described for eastern Australia (Pye 1983, Thompson 1983). Although the causes of later Holocene dune activity may be local, as Pye (1983) has suggested for Queensland, it is tempting to consider wider ranging explanations, especially in view of the evidence now emerging concerning sea-level changes and coastal instability over the past 2000 years.

The four dune phases share several similarities, notably orientations and internal structures indicative of persistent northwest winds; dominant grain-size modality in the 2-3 phi range; roundness and sphericity values which, though varying with grain size, are similar for all dune sands; a paucity of rounded grains suggestive of limited abrasion; and a common detrital mineralogy. These parameters have not changed much with time and must, therefore, be properties of the original deposits and their formative conditions.

Despite these similarities, significant differences permit discrimination between dune phases on the basis of age. Surficial characteristics which change progressively with time include dune morphology which becomes subdued, vegetation character which becomes more complex, and plant density and cover which increase with age. Within the dunes, the grain-size parameters of sorting, skewness and kurtosis show significant changes over time, and multivariate discriminant analysis proves to be an effective tool for distinguishing between dune phases using four textural parameters simultaneously. Scanning electron microscopy reveals that, while most physical microfeatures of quartz grains are inherited from the original aeolian environment, chemical features change significantly over time. This is further explained by mineralogical changes, specifically the solution and reprecipitation of silica, iron and carbonate minerals, and the translocation of clay minerals and aeolian dust through the weathering profile. Over the latitudinal extent of the study, significant regional differences occur within individual dune phases, but the parameters discussed here as discriminating betwen dune phases have wide application.

Acknowledgements

The authors gratefully acknowledge the assistance of Amalie Jo Orme with cartography, Ronald I. Dorn with scanning electron microscopy, and Kevin R. Mulligan with theoretical considerations.

References

Al-Saleh, S. and F.I. Khalaf 1982. Surface textures of quartz grains from various Recent sedimentary environments in Kuwait. *Journal of Sedimentary Petrology* 52, 215-25.

Bowen, A.J. and D.L. Inman 1966. Budget of littoral sands in the vicinity of Point Arguello, California. *Coastal Engineering Research Center, Corps of Engineers, United States Army, Technical Memorandum 19.*

Bradley, W.C. and G.B. Griggs 1976. Form, genesis, and deformation of central California wave-cut platforms. *Geological Society of America Bulletin* 87, 433-49.

Cooper, W.S. 1967. Coastal dunes of California. *Geological Society of America Memoir 104*, Boulder.

Dickinson, W.R. 1981. Plate tectonics and the continental margin of California. In *The geotectonic development of California*, W.G. Ernst (ed.), 1-28. Englewood Cliffs, NJ: Prentice-Hall.

Folk, R.L. and W.C. Ward 1957. Brazos River bar: a study in the significance of grain-size parameters. *Journal of Sedimentary Petrology* 27, 3-26.

Gardner, R. and K. Pye 1981. Nature, origin and palaeoenvironmental significance of red coastal and desert dune sands. *Progress in Physical Geography* 5, 514-34.

Gorsline, D.S. and R.A. Stewart 1962. Benthic marine exploration of Bahía de San Quintín, Baja California, 1960-61, marine and Quaternary geology. *Pacific Naturalist* 3, 282-319.

Goudie, A.S. and A. Watson 1981. The shape of desert dune grains. *Journal of Arid Environments* 4, 185-90.

Greenwood, B. 1969. Sediment parameters and environment discrimination: an application of multivariate statistics. *Canadian Journal of Earth Science* 6, 1347-58.

Gribi, E.A. 1963. Monroe Swell oil field, Monterey County, California. In *Guidebook to the geology of the Salinas valley and San Andreas fault*, American Association of Petroleum Geologists and Society of Economic Paleontologists and Mineralogists, Pacific Section, Annual Spring Field Trip, 76-77.

Hall, C.A. 1978. Origin and development of the Lompoc-Santa Maria pull-apart basin and its relation to the San Simeon-Hosgri strike-slip fault, western California. *California Division of Mines and Geology, Special Report 137*, 25-31.

Hertlein, L.G. and E.C. Allison 1959. Pliocene marine deposits in northwestern Baja California, Mexico, with the description of a new species of *Acanthina* (Gastropoda). *Bulletin of the Southern California Academy of Sciences* 58, 17-26.

Hobday, D.K. and A.R. Orme 1975. The Port Durnford Formation: a major Pleistocene barrier-lagoon complex along the Zululand coast. *Transactions of the Geological Society of South Africa* 77, 141-9.

Howell, D.G. and J.G. Vedder 1981. Structural implications of stratigraphic discontinuities across the Southern California Borderland. In *The geotectonic development of California*, W.G. Ernst (ed.), 535-58. Englewood Cliffs, NJ: Prentice-Hall.

Hubbs, C.L., G.S. Bien and H.E. Suess 1960. La Jolla natural radiocarbon measurements. *American Journal of Science Radiocarbon Supplement* 2, 197-223.

Hubbs, C.L., G.S. Bien and H.E. Suess 1962. La Jolla natural radiocarbon measurements II. *Radiocarbon* 4, 204-38.

Hubbs, C.L., G.S. Bien and H.E. Suess 1965. La Jolla natural radiocarbon measurements. IV. *Radiocarbon* 7, 66-117.

Inman, D.L., G.C. Ewing and J.B. Corliss 1966. Coastal sand dunes of Guerrero Negro, Baja California del Sur, Mexico. *Geological Society of America Bulletin* 77, 787-802.

Jawad Ali, A. and R.A. Ali-Ani 1983. Sedimentological and geomorphological study of sand dunes in the western desert of Iraq. *Journal of Arid Environments* 6, 13-32.

Krinsley, D.H. and J.C. Doornkamp 1973. *Atlas of quartz sand grain surface textures.* New York: Cambridge.

Moiola, R.J. and A.B. Spencer 1979. Differentiation of eolian deposits by discriminant analysis. In *A study of global sand seas*, E.D. McKee (ed.), 53-8. *United States Geological Survey Professional Paper* 1052, Washington.

Moiola, R.J., A.B. Spencer and D. Weiser 1974. Differentiation of modern sand bodies by linear discriminant analysis. *Gulf Coast Association of Geological Societies Transactions* 24, 321-6.

Muhs, D.R. 1982. The influence of topography on the spatial variability of soils in Mediterranean climates. In *Space and time in geomorphology*, C.E. Thorn (ed.), 269-84. London: George Allen & Unwin.

Mulligan, K.R. 1985. *The movement of transverse coastal dunes, Pismo Beach, California, 1982-83.* Unpublished MA Thesis, Los Angeles, University of California.

Orme, A.R. 1973. Barrier and lagoon systems along the Zululand coast, South Africa. In *Coastal geomorphology*, D.R. Coates (ed.), 181-217. Boston: George Allen & Unwin.

Orme, A.R. 1980. Marine terraces and Quaternary tectonism, northwest Baja California, Mexico. *Physical Geography* 1, 138-61.

Ortlieb, L. 1978. Reconocimiento de las terrazas marinas cuaternarias en la parte central de Baja California. *Universidad Nationale Autonóma de Mexico, Instituto de Geología Revista* 2, 200-11.

Page, B.M. 1981. The southern Coast Ranges. In *The geotectonic development of California*, W.G. Ernst (ed.), 329-417. Englewood Cliffs, NJ: Prentice-Hall.

Paskoff, R. 1970. *Le Chili semi-aride: recherches géomorphologiques.* Bordeaux: Biscaye Fréres.

Phleger, F.B. and G.C. Ewing 1962. Sedimentology and oceanography of coastal lagoons in Baja California, Mexico. *Geological Society of America Bulletin* 73, 145-82.

Powers, M.C. 1953. A new roundness scale for sedimentary particles. *Journal of Sedimentary Petrology* 23, 117-9.

Pye, K. 1983. Formation and history of Queensland coastal dunes. *Zeitschrift für Geomorphologie* Supplementband 45, 175-204.

Rittenhouse, G. 1943. A visual method of estimating two-dimensional sphericity. *Journal of Sedimentary Petrology* 13, 79-81.

Shepard, F.P. and R. Young 1961. Distinguishing between beach and dune sands. *Journal of Sedimentary Petrology* 31, 196-214.

Thom, B.G., H.A. Pollach and G.M. Bowman 1978. *Holocene age structures of coastal sand barriers in New South Wales, Australia.* Canberra: Royal Military College.

Thompson, C.H. 1983. Development and weathering of large parabolic dune systems along the subtropical coast of eastern Australia. *Zeitschrift für Geomorphologie* Supplementband 45, 205-25.

Wilson, T.A. and J.L. Mero 1966. Economic deposits of the California offshore area. In *Geology of Northern California*, E.H. Bailey (ed.), 343-53. California Division of Mines and Geology, Bulletin 190.

Woods, A.J. 1978. *Marine terraces between Playa El Marron and Morro Santo Domingo, central Baja California.* Unpublished Ph.D. Dissertation, Los Angeles, University of California.

Yeates, R.S. 1978. Neogene acceleration of subsidence rates in southern California. *Geology* 6, 456-60.

10

Origins of low-angle stratification in aeolian deposits

Gary Kocurek

Abstract

A significant portion of the aeolian rock record is not high-angle cross-strata, but rather consists of low-angle strata (≤ 15 degrees). Possible origins for these strata include interdune, sand-sheet, and apron and plinth deposits. Ideally these deposits are distinct, but preservation potential and conflicting models of bedform migration plus similarities in processes complicate interpretations. The geometries of *interdune deposits* are dune-specific. For transverse crescentic dunes, lensoidal interdune bodies occur climbing in the transport direction. Linear dunes, if envisioned as extending downwind or vertically accreting, produce interdune lenses elongate with the dunes. Recent work suggesting that linear dunes migrate laterally, however, suggests extensive, climbing, sheet-like interdune deposits. Similarly, stationary and vertically accreting star dunes give rise to thick, irregular interdune lenses. In contrast, a net lateral migration for star dunes results in climbing sheets of interdune deposits. Interdune deposits, ideally overlying first-order bounding surfaces and alternating vertically with dune deposits, can also occur amalgamated when a very low angle of climb occurs or can be confused with extra-erg deposits that overlie super bounding surfaces. *Sand sheets* typically occur on the margins of ergs and their deposits mark the basal, upper or lateral fringe portions of aeolian units. Regional extensiveness and a lack of symmetrical sequences of structures can distinguish sand-sheet deposits from some interdune accumulations. Deposits of *dune aprons* and *lower plinths* may occur as the sole record where low net preservation occurs. Plinths and aprons of wind-ripple laminae are best developed on oblique crescentic, linear and star dunes where along-slope secondary airflow is significant. These dunes may leave a record of truncating subsets of low-angle wind-ripple laminae that show a distinct parallelism within subsets.

Introduction

A significant portion of the aeolian rock record does not consist of high-angle cross-stratification, but rather is composed of low-angle strata (for this discussion, 0 to about 15 degrees). A listing of aeolian units in which low-angle stratification is present would probably include every aeolian deposit and range from where such stratification is a minor component to those where it comprises most of the unit.

There are several possible origins for low-angle stratification in aeolian units. Interdune areas and sand sheets are environments where low-angle strata can be deposited. Yet some types of dunes can leave a record consisting entirely of low-angle wind-ripple laminae. Extra-erg deposits may occur interbedded within erg deposits. Moreover, the sequence of processes and events that ultimately result in the rock record of ergs can further complicate the interpretation.

The purposes of this paper is to outline the possible origins of low-angle strata in aeolian units and to evaluate whether or not these can be distinguished. The danger is to routinely interpret such deposits as, for example, interdune deposits, without a careful consideration of other possible origins. Principal aspects of the discussion are outlined in Table 1.

Interdune deposits

Occurrence

In some ergs, interdune areas constitute more of the depositional area than the dunes themselves, and it is well recognized that interdune areas can be sites of deposition and leave a rock record (e.g. Ahlbrandt & Fryberger 1981, Hunter 1981, Kocurek 1981). Different dune types have associated distinctive shapes of interdune areas.

Stratigraphic position

It is something of a truism to state that there cannot be interdune areas without dunes. Yet this is an important realization in the initial recognition of interdune deposits. The vertical alternation of flat-bedded deposits and dune cross-strata over many meters is good *circumstantial* evidence that the flat deposits are interdune in origin. This would arise with the "orderly" migration of dunes and interdune areas with deposition and climb (Rubin & Hunter 1982). In the terminology of Brookfield (1977), interdune deposits would be the flat strata overlying first-order bounding surfaces.

The above argument is *not* to say that there cannot be interdune deposits without dune deposits. Simpson and Loope (1985) demonstrate *amalgamated* interdune deposits without intervening dune deposits. This recognition follows ideas by Rubin and Hunter

Table 1 Summary of low-angle deposits in eolian units.

	Interdune Areas	Extra-erg	Sand Sheets	Dunes
1. Occurrence	– areas between dunes – shape characteristic of dune type	– marginal to ergs – termination of ergs or migration with transgressions, climatic changes, or erg basin restructuring	– margins or erg, some internally or independent initial phase of erg development – erg trailing or leading margin – restructuring of erg	– aprons, plinths, parts of dunes not marked by slipface – grainfall apron on small transverse crescentic dunes – wind-ripple apron or plinth on oblique crescentic dunes – wind-ripple plinths on linear and star dunes
2. Stratigraphic position of depositis	– vertical alternation with dune deposits – overlying first-order bounding surfaces – amalgamated	– lateral fringes of eolian units – within dune/interdune sequences – association with super bounding surfaces	– basal, upper, and lateral fringe portions of eolian units	– preserved solely with low angle of bedform climb or low net preservation of original dune with plinth or apron – amalgamated apron or lower plinth deposits
3. Geometry of deposits	– veneer to few meters (?) in thickness – shape determined by interdune area shape and propagation with time	– veneer to several meters or more (?) in thickness – sheet deposits	– veneer to several meters or more (?) in thickness – blanket or sheet deposits	– function of dune shape and propagation with time in subsets
4. Sedimentary structures	– dry-, damp-, wet-surface deposits – sandstone. mudstone carbonate, evaporite – systematic and symmetrical lateral distribution	– can be similar to interdune and sand-sheet depositis – soil zones, vegetation, burrowing, marine-sabkha depositis, water-table horizons, lag surfaces, relict dune topography	– similar to interdune deposits, but lack systematic and symmetrical variations	– wind-ripple laminae, grainfall – internal laminal parallelism within subsets

(1982) that only a small percentage of the original dune height is typically preserved, and by Kocurek (1981) that interdune deposits are selectively preserved over dune deposits because they occupy the base of the dune/interdune depositional couplet. Hence an interdune origin for flat deposits cannot be ruled out just because dune deposits are not present.

Alternatively, not all flat-bedded deposits occurring between dune strata are interdune deposits. Blakey and Middleton (1983), Driese and Dott (1984) and Loope (1984a) document examples of pronounced intertonguing of erg and adjacent environments (continental, coastal or marine) that results in the interbedding of extra-erg and dune/interdune sequences. Such shifts in the erg boundary could arise, for example, with changes in sea level, sand supply or tectonic pulses in the erg basin or adjacent uplands.

On a broader scale, truncational surfaces that may be overlain by a thin layer of flat strata can represent a major restructuring of an erg, termination of an erg "event" such as with a climatic change, or migration of the entire erg (see review by Kocurek & Oakes in press). Subsequent erg initiation or migration of an erg into the area could result in overlying dune/interdune deposits. Loope (1985), Kocurek and Oakes (in press) and Kocurek and Hunter (in press) describe such examples. Regional or super bounding surfaces (terms coined by Talbot) truncate entire erg sequences and should commonly be at an angular discordance with the climbing first-order bounding surfaces (Loope 1984b, Rubin & Hunter 1984, Kocurek & Oakes in press).

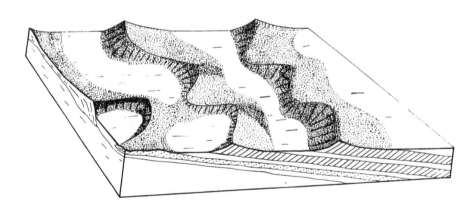

Figure 1 Transverse crescentic dunes and interdune areas showing deposits formed with migration and climb. Angle of climb exaggerated for illustrative purposes. Solid foresets indicate grainflow cross-strata, dashed foresets indicate grainfall or wind-ripple strata.

Geometry of the deposits

The geometry of an interdune deposit is a function of both the *shape* of the interdune area and its *migration* with time. For this discussion, the major basic morphological types of dunes are crescentic, linear and star dunes. In addition, to simplify the discussion and because the shapes of the interdune areas may not be appreciably affected, no discussion is presented of compound or complex dunes.

The basic shapes of interdune areas for the three dune types are evident from inspection of aerial photos of ergs: (a) roughly ellipsoidal, commonly fully enclosed areas elongate parallel to the dune crestline for crescentic dunes, (b) very long corridors trending parallel to the dune crestline for linear dunes, and (c) more varying, enclosed to interconnected, equidimensional to irregular areas for star dunes (Figs 1, 2, & 3).

The migration habits of these types of dunes and interdune areas with time are not entirely understood. As an end-member, the migration of perfectly transverse crescentic dunes and their interdune areas is normal to the dune crestline. Interdune deposits, therefore, consist of generally continuous, climbing layers in outcrops that trend parallel to the migration direction, but are broadly lensoidal (over hundreds of meters) in outcrops that trend perpendicular to the migration direction (Fig. 1) (see Kocurek 1981).

For crescentic dunes that are oriented oblique to the resultant wind direction, a longitudinal component of interdune area migration should occur (Rubin, written communication 1985). This will "extend" the lensoidal geometry of the interdune deposits in a direction more parallel to the dune crestline.

Far more controversial are the migration habits of linear dunes and their interdune areas in part because of the uncertainties concerning the wind regimes they occur in and their orientations to these winds (see discussions in Fryberger 1979, McKee 1979, 1982, Lancaster 1982, Tsoar 1982, 1983, Rubin & Hunter 1985). Presuming, however, that linear dunes and their interdune areas leave deposits, very different sorts of interdune geometries are inferred by conflicting models of linear dune migration. Basically, do linear dunes (a) extend downwind roughly parallel to their crestline without significant lateral motion, (b) remain essentially stationary and vertically accrete, or (c) undergo net lateral migration?

In the first two cases, best described in Bagnold (1954), McKee and Tibbitts (1964) and Tsoar (1982, 1983), the geometry of the interdune deposits are practically identical. Interdune lenses with the same widths as the original interdune areas occur in cross-sections normal to the dune crestline, and continuous layers reflecting the extensive original length *or* the net deposit of the time-transgressive interdune corridor are seen in cross-sections parallel to the crestline (Fig. 2A). This case would also largely apply to linear dunes and their interdune areas that are "frozen" in place by some event such as stabilization by vegetation with subsequent burial or burial beneath deposits of a low-energy marine transgression (see Tanner 1965, Steele 1982, Eschner & Kocurek in press).

The model of linear dunes that undergo net lateral migration is argued by Rubin and Hunter (1985) who suggest that linear dunes probably have some dune-normal component of transport and thus migrate laterally. Here, interdune deposits should consist of extensive layers in cross-sections of any orientation and climb in the direction of lateral transport (Fig. 2B).

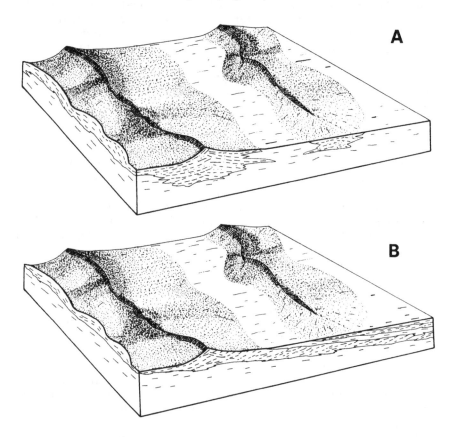

Figure 2 Linear dunes and interdune areas showing deposits. 2A Deposit geometries suggested to occur with lateral migration. Angle of climb exaggerated. Solid foresets indicate grainflow cross-strata, dashed foresets indicate wind-ripple or grainfall strata.

With regard to star dunes, which form in multidirectional wind regimes (Fryberger 1979), Nielson and Kocurek (in review) argue that a slow net migration must occur except in the rare case where all wind components exactly cancel. Earlier, Wilson (1972) argued similarly. This is in contrast to the widely held, but undocumented view that star dunes are stationary and only vertically accrete.

Interdune geometry varies accordingly. Vertically accreting, stationary star dunes preserved within a subsiding basin should be accompanied by thick, relatively isolated to extensive lenses of interdune deposits (Fig. 3A) (see Ahlbrandt & Fryberger 1981). Laterally migrating star dunes, in contrast, should be associated with irregular sheet-like interdune deposits with a preferred direction of climb (Fig. 3B).

The above discussion of the interdune geometries associated with star and linear dunes is largely generated on models alone. Deposits of linear dunes have been only rarely recognized in the rock record (see Rubin & Hunter 1985) and star dune deposits have not been recognized. These models, however, constitute more than an academic exercise in that they imply deposits not currently expected for these dune types (i.e. these deposits are being misidentified in the rock record).

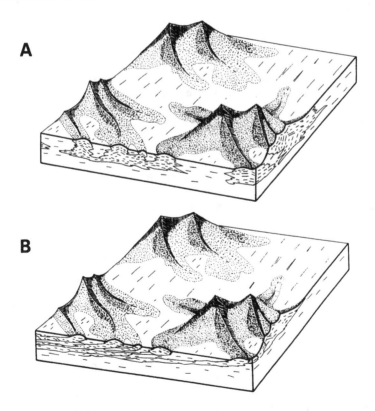

Figure 3 Star dunes and interdune areas showing deposits. 3A Deposit geometries suggested to occur with stationary vertical accretion. 3B Deposit geometries suggested to occur with lateral migration. Angle of climb exaggerated. Solid foresets indicate grainflow cross-strata, dashed foresets indicate wind-ripple or grainfall strata.

Sedimentary structures

Detailed work with both modern and ancient interdune deposits (e.g. Ahlbrandt & Fryberger 1981, Hunter 1981, Kocurek 1981, Hummel & Kocurek 1984, Driese 1985) show that a wide variety of sedimentary structures and lithologies characterize interdune deposits. These include structures found on wet, damp and dry surfaces, and sandstone, mudstone, carbonate and evaporite lithologies. Deposits of individual interdune areas vary in thickness from a few centimeters or less to perhaps a meter or more, with the thickest deposits generally consisting of nonaeolian deposits (Hummel & Kocurek 1984).

Structures and lithologies of interdune deposits can be identical to those of sand sheets. Interdune deposits, however, commonly show a systematic and symmetrical distribution of features away from the center of the interdune area (see outcrop examples in Kocurek 1981, Driese 1985, and modern examples in Hummel & Kocurek 1984), and this is not typical of sand-sheet deposits (Kocurek & Nielson in press a). The most difficult case would be to distinguish mutually truncating, amalgamated interdune deposits from sand-sheet deposits. Here any lateral variations in the sequences of structures might be difficult to discern because of truncation. Determination of an interdune origin would have to rely on the identification of the first-order bounding surfaces that truncate deposits of individual interdune areas.

Sand sheets

Occurrence

Sand sheets are areas of aeolian sands where dunes with slipfaces are generally absent. Sand sheets are present in most modern ergs and have been recognized in the rock record (see review in Kocurek & Nielson in press a). Sand sheets commonly occur along the margins of ergs, although some occur within or independent of a dune field.

Stratigraphic position

The common location of sand sheets along the margins of ergs suggests that their deposits should occupy the basal, upper or lateral portions of aeolian units. This seems borne out by the ancient examples thus far described (Kocurek & Nielson in press a). The occurrence of sand-sheet deposits at the base or top of an aeolian sequence, therefore, can arise with migration of the ergs, although basal sand-sheet deposits may also represent the initial phase of erg development. The occurrence of low-angle deposits at the base of an aeolian unit is not, however, a guarantee of a sand-sheet origin. Initial dune deposits may occur with very little net preservation and the resultant low-angle deposits represent only the lowermost apron or bottomset deposits (i.e. Driese & Dott 1984).

Kocurek and Nielson (in press a) present evidence that sand sheets are caused by factors that act to inhibit dune development. If this is correct, then the occurrence of sand-sheet deposits well within a dune-interdune sequence may signal a major restructuring of the erg. It is conceivable, however, that sand-sheet deposits can occur between sequences of erg sequences and represent the movement of separate ergs across a subsiding basin over a long period of time. In this case the sand-sheet deposits would represent the trailing or leading margins of migrating ergs.

Geometry of the deposit
Sand-sheet deposits should show sheet-like geometries, generally span at least several kilometers and range in thickness from thin veneers to several meters. This size and geometry separates them from some or most interdune deposits.

Sedimentary structures
Kocurek and Nielson (in press a) cite a high water table, surface cementation or binding, periodic flooding, a grain size coarser than that typical of dunes with slipfaces, and vegetation as factors promoting sand-sheet development. These authors argue that one or more of these factors should be evidenced in the accessory structures found in sand-sheet deposits.

Dunes

Occurrence of low-angle strata
Low-angle strata occur as dune apron deposits at the base of a slipface and on the parts of dunes where a slipface is not developed such as the plinths of linear and star dunes or the horns of barchanoid dunes. In the last case, deposits may be laterally traceable into high-angle cross-strata of the slipface, so these sorts of strata are omitted from this discussion. Also omitted here are dune deposits formed in a wind regime where the primary winds associated with slipface advance are punctuated by periods when the primary winds blow along or up the lee face, thus producing alternating packages of wind-ripple strata and grainflow cross-strata extending to the base of the set (see Hunter 1981). What is important here is where slipface deposits (grainflow cross-strata) do not reach the base of the dune and, in some cases, are restricted to the upper portions of the dune.

A large perfectly transverse dune generally does not have a significant apron. This is the result of a well defined break in slope at the dune brink beyond which grains blown onto the lee-face as grainfall do not reach the base of the dune except during extremely high wind velocities, and the back-eddy roller is generally too weak to transport sand from the interdune area back onto the slipface (see Hunter 1981). Apron deposits become more pronounced with small, transverse dunes where grainfall accumulates at the base of the slipface (Hunter 1981).

Large plinths consisting of wind-ripple laminae are best developed where the interactions of the dune and the primary wind cause a secondary airflow that results in transport along the lee face. Such along-slope secondary airflow develops where the dune crestline is not normal to the primary wind, as has been shown for linear dunes (Tsoar 1978, 1982, 1983, Tsoar & Yaalon 1983, Rubin & Hunter 1985), star dunes (Nielson & Kocurek in review), and crescentic dunes that are oblique to the wind (Rubin & Hunter 1983). Indeed, it is apparent that dune deposits characterized by thick basal wind-ripple deposits cannot represent transverse dunes, but rather represent linear, star or crescentic dunes that were not perfectly transverse to the primary wind.

Figure 4 shows a simple linear dune. Mapping of the surface processes on this and other linear dunes in the vicinity shows that most of the surfaces of these dunes are covered by wind ripples and that only the upper 10% is slipface, although some slipfaces extend to the bases of the dunes (Nielson in preparation). Surface measurements and trenching by Nielson show the surface slope as well as the internal cross-strata (wind-ripple laminae) to range from 0-7° near the dune base to 13-16° at mid-height on the plinth to 20-25° just below the base of the slipface. Similarly, low slopes or internal dip angles are shown for the lower portions of linear dunes in trenches by Mckee and Tibbitts (1964) and profiles by Lancaster (1982), Mckee (1982), and Tsoar (1982).

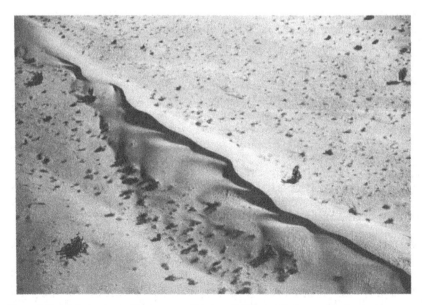

Figure 4 Simple linear dune, Algodones, southeastern California. Slipface confined to uppermost portion of dune. Most of dune shows low-angle surfaces marked by wind ripples.

The morphodynamics of star dunes are also apt to generate extensive plinth deposits (Fig. 5). Nielson and Kocurek (in review) consider star dunes to consist of elements (arms) that are transverse, oblique and longitudinal to any wind, with arms kept active simultaneously by secondary airflow generated on the dune. Because of this and reversals of the slipfaces, many star dunes show slipfaces confined to the upper portion of the dunes with the remaining surfaces marked by wind ripples and underlain by low-angle wind-ripple deposits (see surface maps and trenches in Nielson & Kocurek in review).

Figure 5 Portion of star dune at Dumont, California. High-angle slipfaces occupy upper portion of the dune and rest atop a plinth consisting of wind-ripple deposits.

Stratigraphic position
For any migrating crescentic dunes with an apron where the preserved part of the dune is equal to or less than the apron thickness, low-angle deposits are the only record of the dune. Work thus far (e.g. Kocurek 1981, Rubin & Hunter 1982, 1983, Kocurek & Nielson in press b) suggests that ordinarily the angle of bedform climb is small and the preserved part of the dune is minimal. This suggests that preservation of just the apron deposits of dunes should not be uncommon, and this could result in the vertical stacking of apron deposits with a total absence of slipface deposits. Examples are presented in Driese and Dott (1984).

Simple linear dunes may well be bedforms that largely deposit low-angle wind-ripple deposits. Linear dunes with some component of lateral migration should leave predominantly low-angle wind-ripple deposits unless the angle of climb is unrealistically high (Fig. 2B).Less certain would be the case for linear dunes that remain stationary and vertically accrete along with surrounding interdune areas (Fig. 2A). It is not known whether plinth deposits pass laterally into a core of slipface deposits or if the slipface is continuously displaced upward and underlain by lower angle deposits such that growth of the dune is accomplished by plinth accretion. A linear dune that only extends parallel to its crest because of secondary longitudinal airflow, and with low net preservation, should leave a relatively narrow trail consisting of wind-ripple deposits and grainflow cross-strata where the slipface extended to near the base of the dune (Fig. 2A) (see Tsoar 1982, 1983).

Similarly, laterally migrating star dunes should deposit largely low-angle, wind-ripple deposits. Vertically accreting star dunes and interdune areas may yield more slipface deposits (Fig. 3A). Here too, the critical question is whether or not star dunes maintain their height within a subsiding, infilling basin by the buildup of the plinth.

Geometry of the deposits and sedimentary structures
As apron or plinth deposits originate on dunes, their geometries are a function of the dune shape and propagation behavior. Apron deposits for a perfectly transverse crescentic dune would extend laterally the width of the dune, and extend in the dune propagation direction for as far as the dune migrated and left a deposit. These apron deposits, if they solely form the rock record, would form *sets* of low-angle cross-strata and be separated by a lower bounding surface towards which laminae would gradationally toe, and by an upper truncational bounding surface.

The plinth deposits of linear and star dunes show packages of wind-ripple laminae separated by very low-angle truncating surfaces (Nielson & Kocurek in review, Nielson in preparation). Laminae conform to the lower bounding surface and are truncated at a very low angle by the upper surface. This reflects the shifting of the wind, first forming a reactivation surface and redefining the plinth depositional surface, followed by deposition conforming to this surface (Fig. 6). Laminae within packages or subsets separated by reactivation surfaces generally show a marked *parallelism*, but differ slightly in their orientation from laminae in subsets above or below. These characteristics seem to be diagnostic of plinth or apron deposits (Fig. 7). Laminar parallelism reflects the shape of the depositional surface, which in turn is aerodynamically controlled. This surface changes only as the dune changes shape to some degree. Such aerodynamically controlled depositional surfaces are not common in dry interdune areas or sand sheets and, hence, these deposits do not show such parallelism (Kocurek & Nielson in

Figure 6 Trench in plinth deposits of a linear dune at the Algondones. Deposits are marked by subsets of low-angle wind-ripple laminae separated by low truncational surfaces (arrows). Laminae within subsets tend to be parallel, but at an angular discordance with laminae within other subsets. Vertical height is about 90 cm.

press a). Exceptions here of course would be deposits of zibars, "interdune dunes" and shadow dunes, all of which are aerodynamically controlled.

An additional aspect of low-angle deposits of aprons or plinths may be a steepening of laminae upward in a set or subset, especially as net preservation of the dune increases. This would reflect the steepening of aprons or plinths upward and the tendency of the deposits to reflect the depositional surface.

Conclusions

(a) A significant portion of the aeolian rock record consists of low-angle cross-strata. These can represent deposits of interdune areas, sand sheets and aprons and lower plinths of dunes.

Figure 7 Low-angle wind-ripple laminae, Page Sandstone, Page, Arizona. These deposits are interpreted to represent dune aprons or plinths. Note low-truncating surfaces (arrows) separating subsets of wind-ripple laminae, which show an internal parallelism.

(b) Characteristics of interdune deposits include: (i) regular interbedding with dune strata, (ii) a shape determined by dune/interdune type and migration behavior, (iii) a variety of sedimentary structures that may show systematic and symmetrical sequences when traced vertically and laterally, and (iv) association with first-order bounding surfaces. Problems in identifying interdune deposits arise when they are amalgamated, and in recognizing extra-erg deposits or deposits associated with erg degradation, both of which overlie super bounding surfaces within dune/interdune sequences.

(c) Sand-sheet deposits most typically occur at the basal, upper or lateral fringes of erg units. Sand-sheet deposits are commonly extensive blanket deposits, differing in size and shape from interdune deposits. A suite of sedimentary structures similar to interdune deposits occurs, but these generally lack the vertical and lateral systematic variations of interdune deposits.

(d) Dunes can produce low-angle strata exclusively when apron or lower plinth deposits are solely preserved. Apron strata consisting primarily of grainfall deposits represent aprons on small, perfectly transverse dunes or possibly larger transverse dunes formed at very high wind velocities. Low-angle deposits consisting entirely of wind-ripple laminae are best developed on oblique crescentic, linear and star dunes where secondary airflow produces much along-slope wind. Simple linear and star dunes are apt to leave a record consisting largely of low-angle wind-ripple deposits because the slipfaces are typically confined to the upper portions of the dunes. Dune apron or plinth deposits, because they originate on dunes and are aerodynamically controlled, occur in subsets with a marked internal parallelism of laminae within a subset, and may show an upward increase in dip angle with greater preservation of original dune height.

Acknowledgements

This paper was improved by comments by J. Nielson (University of Texas) and D. Rubin (U.S.G.S., Menlo Park).

References

Ahlbrandt, T.S. and S.G. Fryberger 1981. Sedimentary features and significance of interdune deposits. In *Recent and ancient nonmarine depositional environments: models for exploration*, F.G. Ethridge and R.M. Flores (eds.), 293-314. Special Publication of the Society of Economic Paleontologists and Mineralogists.

Bagnold, R.A. 1954. *The physics of blown sand and desert dunes.* London: Methuen.

Blakey, R.C. and L.T. Middleton 1983. Permian shoreline eolian complex in central Arizona: dune changes in response to cyclic sealevel changes. In *Eolian sediments and processes*, M.E. Brookfield and T.S. Ahlbrandt (eds.), 551-81. Developments in Sedimentology 38.

Brookfield, M.E. 1977. The origin of bounding surfaces in ancient aeolian sandstones. *Sedimentology* 24, 303-32.

Driese, S.G. 1985. Interdune pond carbonates, Weber Sandstone (Pennsylvanian-Permian), northern Utah and Colorado. *Journal of Sedimentary Petrology* 55, 187-95.

Driese, S.G. and R.H. Dott, Jr. 1984. Model for sandstone-carbonate "cyclothems" based upon the Upper Morgan Formation (Pennsylvanian) of northern Utah and Colorado. *Bulletin of the American Association of Petroleum Geologists* 58, 574-97.

Eschner, T.B. and G. Kocurek. In press. Marine destruction of eolian sand seas: origin of mass flows. *Journal of Sedimentary Petrology.*

Fryberger, S.G. 1979. Dune forms and wind regime. In *A study of*

global sand seas, E.D. McKee (ed.),137-69.   U.S. Geological
Survey Professional Paper 1052.

Hummel, G. and G. Kocurek 1984. Interdune areas of the
back-island dune field, north Padre Island, Texas. *Sedimentary
Geology* 39, 1-26.

Hunter, R.E. 1981. Stratification styles in eolian sandstones: some
Pennsylvania to Jurassic examples from the western interior
U.S.A. In *Recent and ancient nonmarine depositional environment:
models for exploration*, F.G. Ethridge and R.M. Flores, (eds.),
315-29. Special Publication of the Society of Economic
Paleontologists and Mineralogists.

Kocurek, G. 1981. Significance of interdune deposits and bounding
surfaces in aeolian dune sands. *Sedimentology* 28, 753-80.

Kocurek, G. and R.E. Hunter. In press. Origin of polygonal fractures
in sand, uppermost Navajo and Page Sandstones, Page, Arizona.
Journal of Sedimentary Petrology.

Kocurek,G. and J. Nielson. In press a. Conditions favorable for the
formation of warm-climate eolian sand sheets. *Sedimentology*.

Kocurek, G. and J. Nielson. In press b. Determination of ancient
eolian dune height from grainflow cross-stratum thickness.
Journal of Sedimentary Petrology.

Kocurek, G. and C. Oakes. In press. Migration of dunes and
ergs-bounding surfaces revisted. In *Late Paleozoic and Mesozoic
eolian systems in the western U.S.*, G.A. Kocurek (ed.), Special
Publication of the Society of Economic Paleontologists and
Mineralogists.

Lancaster, N. 1982. Linear dunes. *Progress in Physical Geography* 6,
476-504.

Loope, D.B. 1984a. Eolian origin of Upper Paleozoic sandstones,
south-eastern Utah. *Journal of Sedimentary Petrology* 54, 563-80.

Loope, D.B. 1984b. Origin of extensive bedding planes in aeolian
sandstones: a   defense   of   Stokes'   hypothesis--discussion.
Sedimentology 31, 123-5.

Loope, D.B. 1985. Episodic deposition and preservation of eolian
sands: a late Paleozoic example from southeastern Utah. *Geology*
13, 73-6.

McKee, E.D. 1979. Introduction to a study of global sand seas. In *A
study of global sand seas*, E.D. McKee (ed.), 1-19. U.S.
Geological Survey Professional Paper 1052.

McKee, E.D. 1982. Sedimentary structures in dunes of the Namib
desert, Southwest Africa. *Geological Society of America Special
Paper* 188.

Mckee, E.D. and G.C. Tibbitts. 1964. Primary structures of a seif
dune and associated deposits in Libya. *Journal of Sedimentary
Petrology* 34, 5-17.

Nielson, J. In preparation. Linear dunes of the Algodones Dune
Field, southeastern California.

Nielson, J. and G. Kocurek. In review. Development, processes,
migration and deposits of star dunes, Dumont, California.
Geological Society of America Bulletin.

Rubin, D. M. and R. E. Hunter 1982. Bedform climbing in theory and nature. *Sedimentology* 29, 121–38.

Rubin, D.M. and R.E. Hunter 1983. Reconstructing bedform assemblages from compound crossbedding. In *Eolian sediments and processes*, M.E. Brookfield and T.S. Ahlbrandt (eds.),407-27. Developments in Sedimentology 38.

Rubin, D.M. and R.E. Hunter 1984. Origin of first-order bounding surfaces -- reply. *Sedimentology* 31, 128-32.

Rubin, D.M. and R.E. Hunter 1985. Why deposits of longitudinal dunes are rarely recognized in the rock record. *Sedimentology* 32, 147-57.

Simpson, E.L. and D.B. Loope 1985. Amalgamated interdune deposits, White Sands, New Mexico. *Journal of Sedimentary Petrology* 55, 361-5.

Steele, R.P. 1982. Longitudinal draa in the Permian Yellow Sands of northeast England. In *Eolian sediments and processes*, M.E. Brookfield and T.S. Ahlbrandt (eds.), 543-50 . Developments in Sedimentology 38.

Tanner, W.F. 1965. Upper Jurassic paleogeography of the Four Corners area. *Journal of Sedimentary Petrology* 35, 564-74.

Tsoar, H. 1978. *The dynamics of longitudinal dunes*. London: U.S. Army European Research Office.

Tsoar, H. 1982. Internal structure and surface geometry of longitudinal (seif) dunes. *Journal of Sedimentary Petrology* 52, 823-31.

Tsoar, H. 1983. Dynamic processes acting on a longitudinal (seif) sand dune. *Sedimentology* 30, 567-78.

Tsoar, H. and D.H. Yaalon 1983. Deflection of sand movement on a sinuous longitudinal (seif) dune: use of fluorescent dye as tracer. *Sedimentary Geology* 36, 25-39.

Wilson, I.G. 1972. Aeolian bedforms -- their development and origins. *Sedimentology* 19, 173-210.

11

Aeolian landforms: laboratory simulations and field studies

Ronald Greeley

Abstract

Aeolian landforms result from the complex interplay of wind, windblown particles, and the surface. Using results from wind tunnel simulations, field studies (including remote sensing), and theoretical considerations, three landforms have been assessed to determine their aeolian characteristics: domical hills, yardangs, and craters. For both domical hills and raised-rim craters, a horseshoe vortex wraps around the obstruction and forms a distinctive zone of erosion and deposition. The formation of linear lee dunes in the wake of some hills and other topographic obstructions may be the consequence of this wind pattern under some conditions dependent upon sand supply and general wind speed. In the evolution of yardangs, hills are eventually sculpted to a 1:4 width-to-length ratio, which from aerodynamical condition is an optimal streamline form. Future work will involve modelling dynamic dunes (made possible with a new wind tunnel facility) and field studies using multiple arrays of field-portable anemometers to obtain concurrent wind velocity profiles around different landforms.

Introduction

A wide variety of erosional and depositional landforms develop as a consequence of aeolian processes. Ranging in size from centimeter sized ripples to sand-seas covering thousands of square kilometers, aeolian landforms result from the interaction of the atmosphere, lithosphere, and windblown particles. Aeolian features have long intrigued engineers and scientists, as reviewed by Cooke and Warren (1973). With the discovery of aeolian landforms on Mars, consideration of aeolian processes also caught the attention of planetary scientists (reviewed by Greeley & Iversen 1985). In addition, because of the close link with desertification, understanding aeolian processes has cultural and economical implications, especially for agricultural areas being overrun by sand, or undergoing deflation. Various studies are underway to determine how sand movement can be stabilized or otherwise controlled in such areas (Hassan in press). Thus, knowledge of how landforms and other obstacles to the wind influence erosion and deposition is of interest for the most basic of human activities - production of food - to the most sophisticated - the exploration of space.

In this paper, the approach for studying aeolian processes in relation to landforms is discussed and three landforms are analyzed: domical hills, yardangs, and craters. For each of these landforms, consideration is given to the air flow around the objects and the resulting zones of wind erosion and deposition.

Methodology

The approach used in this study combined results obtained from laboratory simulations, field studies (including measurements of wind speeds around various landforms, and analyses of remote sensing data), and considerations of theory. The remote sensing aspects included the study of conventional aerial photographs and images obtained from Earth orbit. Synoptic-scale images from orbit allow assessment of aeolian processes on a regional scale and provide an important basis for the interpretation of extraterrestrial planets.

Field work was critical in this study because it provided an important "check" of the laboratory simulations. One of the most important parameters in field studies of aeolian processes is the wind drag velocity, u_* (Bagnold 1941), also called the "friction speed". This parameter is derived from the wind velocity profile, which typically is obtained from wind speeds measured at several (a minimum of three) heights above the surface. This is commonly accomplished by having anemometers arrayed on a mast, or meteorological tower. In order to characterize the flow field and wind shear distribution around landforms, concurrent wind velocity measurements from many arrays are desirable, but because the cost for instrumentation is high, such measurements have seldom been made. Field work at Amboy Crater, one of the domical hill sites in this study, included the use of three masts, each 16 m high and with five anemometers. Although these measurements were useful to test general models, they were inadequate to characterize the detailed flow field. Consequently, a less expensive (but labor-intensive) system was developed which used simple, non-recording anemometers placed on lightweight masts (Fig. 1). During the experiment period when winds were to be measured, one person was assigned to each station to record the wind speeds and directions as a function of time. The disadvantage of this technique was the need for as many people as there were stations, and the requirement for continuous on-site monitoring for the

Figure 1 Anemometer array used in the yardang study at Rogers Lake, California. View is upwind with the plays, Rogers Lake, in the background. Three anemometers were spaced at heights using a logarithmic scale above the ground on a light-weight mast 5m high. Use of common aluminum tubes (tent-poles), inexpensive anemometers, and standard plastic pipe fittings enabled the purchase of multiple arrays to be deployed to obtain concurrent wind velocity measurements (arrows show locations of some of the other masts); (ASU photograph 2841-D, 1985).

entire time of the experiment. Although not used in the study at the Amboy site, this system was deployed at the yardang locality.

A critical part of the study involved the use of laboratory simulations using wind tunnels to model the landforms of interest. Laboratory simulations have the advantage that conditions, such as wind speed and direction, can be controlled and can provide results that complement field data. Such simulations are often used in engineering applications to address problems of windblown particles (Iversen 1981 and others). Two wind tunnels were used in this study. The first system was an open-circuit atmospheric boundary-layer wind tunnel that is 16 m long and has a 1.25 by 1.25 m cross-section. This tunnel was used to model the wind flow-field as a function of time and landform geometry and to determine zones of wind erosion and deposition. The second system is a "two-dimensional" wind tunnel which emits smoke streams in order to assess the flow field for different landform geometries in profile. It is ideally suited to determine zones of flow attachment, separation, and reattachment, the identification of which are critical for the development of models of aeolian landform evolution.

In the final part of the analysis, results from the field studies and laboratory simulations were compared with theory to derive a general model for the aeolian characteristics of the landforms studied.

Domical hills

Various hills constitute one of the most common types of obstacle to the wind. In some cases, zones of erosion seem to dominate the wake ("down wind" or "lee") zone of hills; in other cases, the wake zone is characterized by extensive sand deposits, as shown by the linear dunes found in the lee of some mountains (Fig. 2). In order to assess aeolian processes around hills, one site was studied in detail, Amboy, California.

Figure 2 Low oblique aerial view of a linear lee dune formed down wind from a small hill on the edge of Pampa Coscomba, Peru. The dune is more than 1 km long, 100m wide, and 30m high (from Grolier et al., 1974).

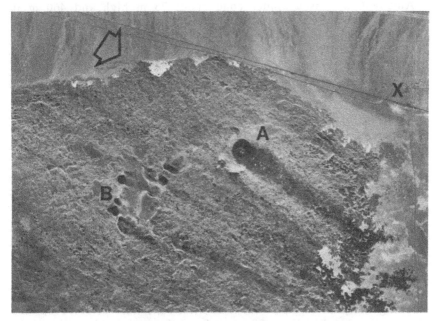

Figure 3A Vertical aerial photographs of the Amboy lava field, showing the prominant dark streak formed in the lee of the cinder cone (A), (B); (x) is the location of Amboy community; area shown is about 11 by 8 km. Arrow indicates prevailing wind direction as measured in the field.

Figure 3B Oblique aerial view toward the west showing the location of meteorological tower no. 1 (to obtain wind velocity profile for winds blowing over smooth plain), tower no. 2 (to measure winds over the rough lava flow), and tower no. 3 (to measure winds in the wake of Amboy cone); (from Greeley and Iversen, 1978).

Amboy Crater is a cinder cone about 75 m high and 460 m in diameter situated in a basaltic lava field on the Mojave Desert of California (Fig. 3A). Sand is blown across the field from the west and drifts around the cone, leaving distinctive "wind streak" patterns when viewed from the air. The most prominant feature is a large dark streak in the wake of the cinder cone. The dark streak could result from: (a) downwind deposits of cinders eroded from the cone, (b) a wind "shadow" zone consisting of low velocity winds into which sand does not move, or (c) a zone of high wind turbulence in which sand is prevented from accumulating. Field inspection (Greeley & Iversen 1978) shows that (a) is not the case because the dark streak lacks cinder deposits. To address (b) and (c), laboratory simulations and field experiments were conducted. Field work included measuring wind velocity profiles inside the wake zone (on the streak) and in an area outside the streak (Fig. 3B). After two months of continuous monitoring (Greeley & Iversen in press), sufficient data were obtained to assess the wind flow for a variety of wind speeds and directions. The wake zone was found to be very turbulent for the northwesterly winds which are primarily responsible for the streak.

The wind flow-field around the cone was also assessed by tracking the path of small balloons. Balloons were released upwind of the cone and their path traced as they were blown past the cone. As they passed the flank of the cone, the balloons were swept into the lee zone. Although most of the balloons broke on the rough lava in the lee zone and the nature of their movement by the wind is not quite the same as saltating sand, qualitatively the experiment suggests that the wind flow would carry sand into the wake zone.

To further assess the flow field around Amboy cone and to gain insight into the associated aeolian processes, a wind tunnel model of the site was constructed at 1/1000 scale (Iversen & Greeley 1978). It included both the cinder cone and the general surface of the hummocky lava flow around the cone (Fig. 4). The model was oriented with its long axis parallel to the prevailing northwesterly winds and to the dark streak. In one experiment, sands were blown across the model and photographs were taken to document zones of accumulation and non-accumulation (Fig. 5). In another experiment, wind velocity profiles were taken in the wind tunnel around the model cinder cone and in its wake to assess the flow field and zones of maximum surface drag. It was determined that the wake zone is very turbulent and that sands drift into the zone, but are swept clear by the turbulence. This supports the hypothesis that the dark streak results from turbulent winds that are shed from the cone and which prevent deposition of sand; thus, the streak is dark because the basaltic lava flow remains free of sand deposits.

The results obtained from the Amboy study, both in the field and in the laboratory, are in good agreement with a flow involving a

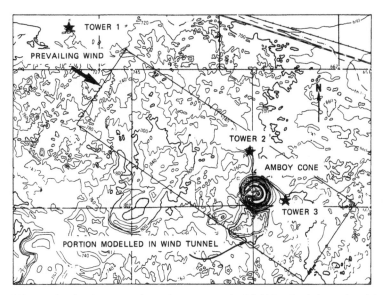

Figure 4 Diagram showing the topography used to construct the 1/1,000-scale wind tunnel model of the Amboy site; long dimension of the model is oriented parallel to the dark streak and the prevailing wind direction.

horseshoe vortex described by Gregory and Walker (1951). For flow around an object having a low height-to-width aspect ration, two trailing vorticies (the "horseshoe") having axes parallel to the wind are shed from the flanks of the object. As applied to landforms (Greeley *et al.* 1974), this results in distinctive zones of high and low wind drag and corresponding areas of erosion and deposition (Fig. 6). Patterns similar to the Amboy streak have been observed in many other places, including a volcanic field east of Damascus, Syria (Fig. 7), in northern Sudan, and on Mars.

Some hills have extensive deposits of sand in their lee (Fig. 2). Although the flow field shown in Figure 6 shows a small zone of deposition in the lee of the obstacle, the length of the zone typically is less than the diameter of the hill or obstacle. How, then, can the very long deposits form? Although no clear answer is in hand, linear lee dunes may form as a consequence of a delicate balance between wind speed and availability of sand. It is well known that once sand accumulates, the deposit often serves as a nucleus for additional sand accretion (Bagnold 1941). The sand deposit in the immediate lee of the obstacle resulting from reverse flow (Fig. 6) could be the nucleus for additional accumulation if the net flux of sand into the zone exceeds the carrying ability of average winds in the area to remove it. Once the deposit begins to grow, it could serve as an ever-lengthening obstacle to the wind, so that the reverse-flow zone is progressively shifted downwind. This hypothesis has been tested tentatively in wind tunnel experiments. In summary, aeolian processes associated with

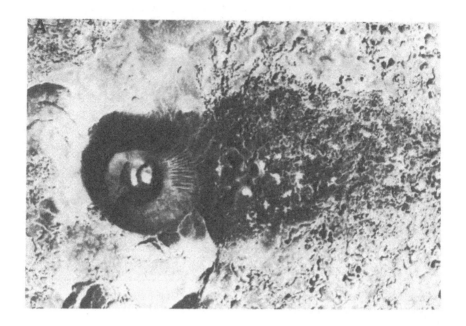

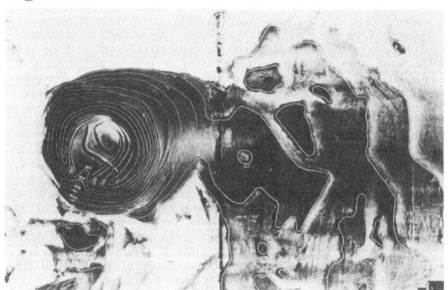

Figure 5 Comparison of a wind tunnel model simulation with the full-scale cinder cone at the Amboy lava field; 5A aerial view of the Amboy crater, wind direction left to right, 5B Iowa State University wind tunnel model of the Amboy crater (scale 1/1,000-scale); 90 μm glass spheres were introduced upstream of the test-section and were then blown across the model. The irregular edge of the dark streak is due to variations in topography for both model and in the field (from Iversen and Greeley, 1978).

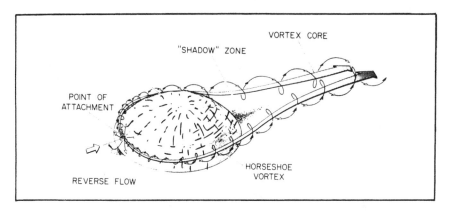

Figure 6 Diagram showing 'horseshoe vortex' formed around a low topographic obstruction. Deposition occurs primarily at the front of the obstruction, in its 'shadow' zone, and on the outside margin of the trailing vorticles. Erosion in pronounced beneath the vortex cores.

hills or similar topographic obstructions involve predominantly wind flow in the form of a horseshoe vortex. The zones of wind erosion and deposition resulting from this flow field may be modified in intensity and geometry, depending upon availability of windblown particles, overall wind speed and direction, the geometry of the landform, and various second-order flow patterns.

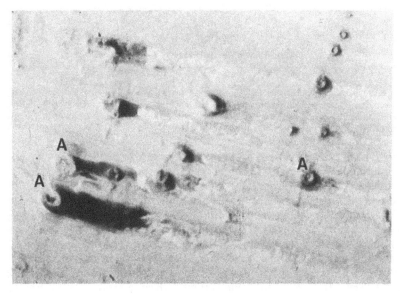

Figure 7 Large-format camera image obtained during an orbital shuttle flight showing a Holocene basalt flow east of Damascus, Syria and wind streak patterns found in the wake of cinder cones (A); most of the lava flow surface is convered by windblown sand; the dark areas are exposed basalt flows; area shown is about 25 km by 18 km; prevailing wind is from the west (left) (courtesy of F. El-Baz, Itek Corporation).

Yardangs

Yardangs are streamlined, aerodynamically-shaped elongate hills oriented parallel to the wind. The term is derived from the Turkistani word *yar* meaning ridge or steep bank from which material is being removed (McCauley *et al.* 1977). First studied by Hedin (1903), recent reviews of yardangs are provided by Mainguet (1972), McCauley *et al.* (1977), and El-Baz *et al.* (1979). Unlike the discussion above which dealt with erosion and deposition associated with hills, in this section it is the object (i.e. the yardang) that is being modified by the wind.

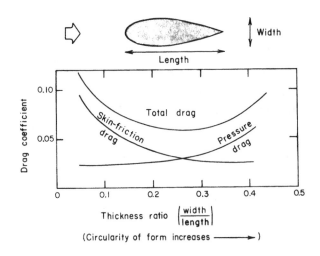

Figure 8 Graph showing coefficient of drag on a streamlined form as a function of the width-to-length ratio, with contributions of skin-friction drag and pressure drag to the toal drag. Open arrow indicates wind direction (from Ward and Greeley, 1984, after Fox and McDonald, 1973).

Ward and Greeley (1984) presented results of field work, wind tunnel modelling, and theoretical analysis of the evolution of yardangs. They showed that yardangs evolve to a 1:4 width-to-length ratio that is consistent with the form expected, based on theory (Fox & McDonald 1973 and Fig. 8). Depending upon the initial shape of the hill, in wind tunnel experiments erosion typically occurred sequentially on: first the windward corners, then the windward slope, then the leeward corners and flanks, and, finally, the leeward slope. Concurrent deposition in the wake of the hill hastened the evolution to a 1:4 width-to-length ratio. Figure 9 shows results from these experiments.

In January 1985, a field experiment was attempted in which arrays of anemometers (Fig. 1) were deployed over one of the yardangs at Blackwelder's (1934) classic locality at Rogers Lake, California. The objective was to obtain concurrent wind velocity measurements over the entire yardang for comparison with the wind tunnel results and theoretical considerations discussed by Ward and Greeley (1984). Unfortunately, winds were inadequate in strength and direction and the experiment was deferred to a later time.

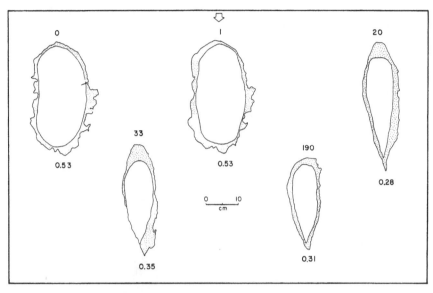

Figure 9 Plan-views showing sequence of yardang evolution in wind tunnel tests, all at a free stream wind speed of 10 m/sec. Numbers at top of each diagram indicate hours from initiation of run, numbers at the bottom indicate the width-to-length ratio; arrow indicates wind direction. Stipple pattern indicates loose particles (after Ward and Greeley, 1984).

Craters

Although craters are not common on Earth, they dominate the martian surface where they frequently display features attributed to the wind. Termed "wind streaks", or "variable features" because they commonly appear, disappear, and change their size, shape, and orientation with time (Sagan *et al.* 1972), they can be used as "wind vanes" to map local near-surface winds. Some wind streaks are considered to be zones of erosion in the wake of craters, whereas others are thought to be deposits of sand or dust. Thus, the problem posed is to determine how craters serve as obstacles to the wind and to assess their influence on aeolian processes.

Depending upon their mode of origin (impact, volcanic, collapse, etc.) and other factors, craters display a wide variety of morphologies. In order to determine the influence of specific geometry on aeolian processes, models of different crater shapes were studied in the wind tunnel. Two types of experiments were run for most shapes: (a) erosion, in which sands were spread uniformly over the model and then run to determine where erosion began and how the patterns evolved, and (b) deposition, in which sands drifted across the model and zones of preferential deposition were noted. From these experiments, a general flow field was derived for raised-rim craters (Greeley *et al.* 1974), the type commonly found on Mars. Zones of erosion and deposition are very similar to those defined for flow around hills, i.e. correspondong to the "horseshoe vortex" flow (Fig. 6).

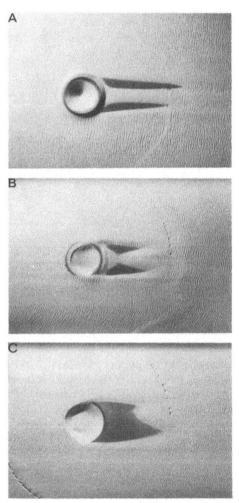

Figure 10 Views of crater model run in the wind tunnel showing erosion patterns as a function of crater rim height. All models were covered with a uniform layer of particles and subjected to the same wind (~380 cm/sec); **10A** high raised rim crater resulted in two distinctive erosional zones; **10B** low raised rim crater resulted in wider erosional zones than in **10A**; and **10C** rimless crater formed a single erosional zone in the crater wake.

From the experiments, crater rim height was found to have a significant influence on the pattern of wind erosion and deposition in the wake of the crater (Fig. 10). As the rim height is decreased, the two zones of erosion shed from the rim flanks tend to merge toward the center, with a corresponding decrease in the size of the depositional zone in the immediate lee of the crater. For a rimless crater (a simple bowl), the lee zone is dominated by erosion (Fig. 10C). As portrayed in Figure 10 and discussed in the section dealing with hills, details of the erosion and deposition pattern are also influenced by many other factors, including general wind speed.

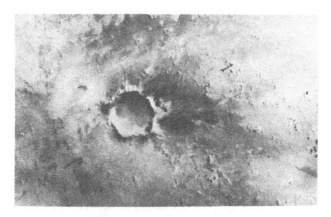

Figure 11 Vertical aerial photograph of Aouelloul Crater, Algeria. Light areas are mostly windblown sand deposits, dark areas are predominantly bedrock exposures free of deposits around this 250m in diameter impact crater; (photograph courtesy of the Smithsonian Inst. Washington, D.C.).

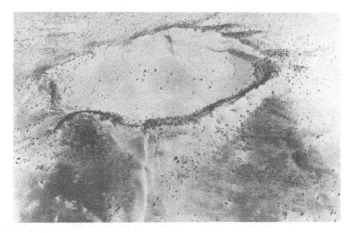

Figure 12 Low oblique aerial photograph of Aouelloul Crater, Algeria, viewed toward the 'up-wind' direction, showing the linear dune in the lee of the crater rim. (photograph courtesy of the Smithsonian Inst , Washington, D.C.).

Although no wind measurements have been made around large, natural craters, the patterns of wind erosion and deposition have been assessed for several craters using remote sensing data. Most of the examples are natural impact craters in desert areas experiencing active sand movement. Figures 11-14 show these craters and the zones of erosion and deposition. Comparison with the wind tunnel experiments shows a close correspondence with the model shown in Figure 6. In particular, Aouelloul crater (Fig. 11) shows the distinctive bilateral symmetry involving twin erosional zones which correspond to the vorticies shed from the rim flank. The linear accumulation of sand in the immediate lee of the crater rim (Fig. 12) is similar in general form to that seen in the wind tunnel.

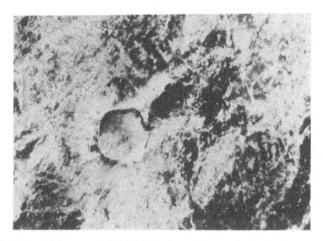

Figure 13 Vertical aerial photograph of Temimichat Impact Crater, Algeria, showing sand deposits (light) and exposed bedrock (dark) (photograph courtesy of the Smithsonian Inst., Washington, D.C.).

Figure 14 shows a small collapse depression on the Amboy lava flow. This rimless crater shows a wind erosion and deposition pattern analogous to that in the wind tunnel run for a rimless crater (Fig. 10C).

Discusssion and future studies

From the results presented here, it is shown that the influence of moderate-scale topography (objects ≤ 5 km across) on aeolian processes can be assessed using a combination of field studies and laboratory simulations to develop predictive models. Wind tunnel simulations have the advantage that conditions can be controlled and are uniform. This permits various parameters to be isolated and studied to determine their role in the overall processes of wind erosion and deposition.

Figure 14 Oblique aerial view of 10-meter-in-diameter collapse crater (foreground) and small dome (upper left) on the Amboy lava field. Prevailing wind is from the northwest (left side of picture). Sand (white areas) collects in the lee of the dome and is the lee of the up-wind rim of the crater.

Until recently, it has been difficult to model sand dune dynamics at wind tunnel scales (Tsoar 1983) . The primary problem is to satisfy the length-ratio similitude parameter. Ideally, the saltation path length-to-dune length ratio should be the same in the wind tunnel as in nature. Thus, proper scaling of dunes in the wind tunnel requires saltation paths to be less than a few millimeters, but sand saltation paths of this length do not occur in air, regardless of particle size, density, or wind speed. However, a recent discovery using a new wind tunnel may now make possible the modelling of dynamic dune forms. The Venus Wind Tunnel was fabricated to simulate aeolian processes on Venus (Greeley *et al.* 1984b). It operates at ambient laboratory temperatures and with a carbon dioxide atmosphere at 35 bars pressure to produce the same atmospheric density as on Venus. Because of the high fluid density, very low threshold wind speeds are adequate for particle entrainment and grains move in very short trajectories (a centimeter or so). From preliminary experiments, it was discovered that small (< 10 cm) dune-like features developed in the Venus Wind Tunnel. Termed *microdunes* (Greeley *et al.* 1984a), these features are asymmetric and have slip faces, display internal bedding, show flow separation from the dune crest and reattachment in the wake, and have a wavelength-to-height ratio of 24:1 to 12:1. All of these characteristics are typical for full-scale sand dunes.   Thus, it appears possible to simulate some aspects of dynamic dunes using the Venus Wind Tunnel and such experiments are planned for the future.

Field studies are critical in the assessment of aeolian processes. Not only do they allow validation of results obtained in the laboratory simulations, but natural conditions are often much more complex than can be modelled.   Several additional field experiments are planned in order to extend the results presented here.   As discussed above, the experiment attempted at Rogers Lake, California, will be repeated to obtain wind measurements over a typical yardang.   Similar experiments are planned to assess wind flow over and around various craters.

In conclusion, aeolian erosion and deposition is a complex process involving interaction among multiple wind speeds and directions, windblown particles, and topography.   Understanding how the process operates and where erosion and deposition are likely to occur can be gained through a combination of laboratory simulations, field studies, and theoretical modelling.

Acknowledgements

This work was supported by the Office of Planetary Geology, National Aeronautics and Space Administration.

References

Bagnold, R.A. 1941. *The physics of blown sand and desert dunes.* Methuen: London.

Blackwelder, E. 1934. Yardangs. *Geological Society America Bulletin* 45, 159-66.

Cooke, R.U. and A. Warren 1973. *Geomorphology in deserts.* Berkeley: University California Press.

El-Baz F., C. S. Breed, M.J. Grolier and J.F. McCauley 1979. Eolian features in the western desert of Egypt and some applications to Mars. *Journal Geophysical Research* 84, 8205-21.

Fox, R. and A.T. McDonald 1973. *Fluid mechanics.* New York: Wiley.

Greeley, R. and J.D. Iversen 1978. Field guide to Amboy lava flow, San Bernadino County, California.   In *Aeolian features of southern California: a comparative planetary geology guidebook,* NASA, 23-52.

Greeley, R. and J.D. Iversen 1985. *Wind as a geological process.* Cambridge: Cambridge University Press.

Greeley, R. and J.D. Iversen In press. Aeolian processes and features at Amboy lava field, California. *Proceedings, UNESCO Workshop on Physics of Desertification.* Trieste, Italy.

Greeley, R., J.R. Marshall and R.N. Leach 1984a. Microdunes and other aeolian bedforms on Venus: wind tunnel simulations. *Icarus*

60, 152-60.

Greeley, R., J.D. Iversen, J.B. Pollack, N. Udovich and B. R. White 1974. Wind tunnel simulations of light and dark streaks on Mars. *Science* 183, 847-49.

Greeley, R., J.D. Iversen, R. Leach, J. Marshall, B. White and S.Williams 1984b. Windblown sand on Venus: preliminary results of laboratory simulations. *Icarus* 57, 112-24.

Gregory, N. and W. Walker 1951. *Great Britain Aeronautics Research Council.* Report Memo No. 2779.

Grolier, M., G.E. Ericksen, J.F. McCauley and E.C. Morris 1974. The desert landforms of Peru: a preliminary photographic atlas, U.S. Geological Survey Interagency report. *Astrogeology* 57, 146 pp.

Hassan, M. In press. The physics of desertification, introduction. *Proceedings, UNESCO Conference.* Khartoum.

Hedin, S. 1903. *Central Asia and Tibet.* Vols. 1 and 2. New York: Charles Scribner and Sons.

Iversen, J.D. 1981. Comparison of wind tunnel model and full scale snow fence drifts. *Journal of Wind Engineering and Industrial Aerodynamics* 8, 231-49.

Iversen, J.D. and R. Greeley 1978. Atmospheric and wind tunnel experiments of the Amboy crater sand-covered lava flow. *Engineering Research Institute Report.* 78235.

Mainguet, M. 1972. *Le modele des gres.* Paris: L'Institut Geographique National.

McCauley, J.R., M.. Grolier and CS. Breed 1977. Yardangs. In *Geomorpholgy in arid regions,* D.O. Doehring (ed.), 233-69. State University, Binghampton: Proceedings, 8th Geomorphology Symposium.

Sagan, C., J. Veverka, P. Fox, R. Dubisch, J. Lederberg, E. Levinthal, L. Quam, R. Tucker, J.B. Pollack and B.A. Smith 1972. Variable features on Mars: preliminary Mariner 9 television results. *Icarus* 17, 346-72.

Tsoar, H. 1983. Wind tunnel modelling of echo and climbing dunes. In *Aeolian sediments and processes,* M.E. Brookfield and T.S. Ahlbandt (eds.), 247-59. Amsterdam: Elsevier.

Ward, A.W. and R. Greeley 1984. The yardangs at Rogers Lake, California. *Geological Society America Bulletin* 95, 829-37.

12

Aeolian processes and landforms in glaciofluvial environments of southeastern Baffin Island, N.W.T., Canada

Cheryl McKenna-Neuman and Robert Gilbert

Abstract

The arctic proglacial sandur is one member in the global variation of aeolian processes and forms for which a general model is not yet available, although idealized models applicable to subtropical and temperate environments have been developed by a number of workers. The environmental control which clearly distinguishes the proglacial setting from others relates to the seasonal influence of moisture as water, ice and snow. Large quantities of glacial meltwater effectively submerge much of the sandur surface during the summer, while throughout the remainder of the year, snow cover and interstitial ice can effectively stop aeolian erosion and transport. However, violent sand storms do occur on the sandurs of Baffin Island during particularly dry, cold winters. The source of sediment for winter transport may be provided by (a) sandur sediments which are progressively dried by evaporation and sublimation and (b) the abrasion of the frozen sandur surface by the impact of other saltating grains. As typical of other sandurs on Baffin Island, extensive aeolian deposits flank the Weasel River sandur in Pangnirtung Pass and the Maktak sandur on eastern Cumberland Peninsula. These include sand sheets, climbing dunes, sand drifts and vegetated accumulations. Topography, surface roughness and vegetation are strong controls over the form of these deposits and as a result, they clearly resemble many of the aeolian deposits described from other warmer environments. Unlike the deflation of the sandur itself which is strongly influenced by seasonal adjustments in moisture, the adjacent aeolian deposits are relatively dry throughout the summer and bare of snow in the winter, and therefore, are subject to reworking year round.

Introduction

While detailed surveys of aeolian landforms and transport rates have been made in subtropical and temperate environments, the understanding of high latitude, cold environments is fragmentary and more qualitative. In Victoria Valley, Antarctica, elongated pebble ridges, sand dunes and sand ridges are described by Selby *et al.* (1973) and ventifact pavements by Adie (1970). Embleton and King (1975) give a general review of the ventifacts and sand and silt deposits found in central North America and Europe which formed by strong off-glacier winds blowing over glacial and glaciofluvial sediments during the Pleistocene.

Nickling's (1978) work is presently the only study in which aeolian processes have been systematically investigated in detail in a periglacial environment. Nickling measured sediment transport in surface creep, saltation and suspension from May through July on a proglacial delta in the Slims River Valley, Yukon Territory. Major dust storm events appeared to be associated with a distinct set of atmospheric conditions.

In contrast, the autumn and winter months appear to be the period of most significant aeolian activity on the sandurs of Baffin Island. Violent winter sand storms carrying sand up to two meters above the ground are described in R.C.M.P. reports (MacGregor 1924). Andrews *et al.* (1979) also observed evidence of deflation from the flood plain in Pangnirtung Pass during this period. Gilbert (1982) and Shaw *et al.* (1978) suggest that aeolian transport onto the frozen surface of the fiord in winter is a significant source of sediment in benthic deposits. Mullen *et al.* (1972) indicate that about 10% of the sediment deposited in the Arctic Ocean is of aeolian origin. Although aeolian transport can occur on the Baffin sandurs in summer (Church 1970), our observations in Pangnirtung Pass indicate that these events are usually small and relatively infrequent. Sand sheets, sand dunes, and interbedded aeolian sand and vegetation are found at the heads of many of the eastern fiords of Baffin Island (Andrews & Dyke 1974, Gilbert 1983, Syvitski *et al.* 1983).

This paper summarizes preliminary investigations on (a) the controls which distinguish proglacial aeolian activity from that in other environments, and (b) the form and stratigraphy of aeolian landforms in eastern Baffin Island.

Study area

Physical setting
The Davis Highlands physiographic region which extends along the east coast of Baffin Island from Bylot Island to Cumberland Peninsula comprises an area of fretted mountains rising 1500 to 2050 m above sea level, surrounded by a dissected plateau having

lower elevation and less relief (Dyke 1979). Less than 5% of the area is covered by permanent ice. Almost one third of Cumberland Peninsula is ice covered including the Penny Ice Cap (6000 km^2), its ten major outlet glaciers, and the smaller ice caps and cirque glaciers to the southeast (Fig. 1). The coast is deeply indented with fiords (Gilbert 1984), many of which are joined by valleys such as Pangnirtung Pass which cut through the mountains. Neoglacial deposits (Davis 1985) and associated glaciofluvial sandurs in these valleys form fresh, unvegetated surfaces which are sources of abundant material for aeolian activity.

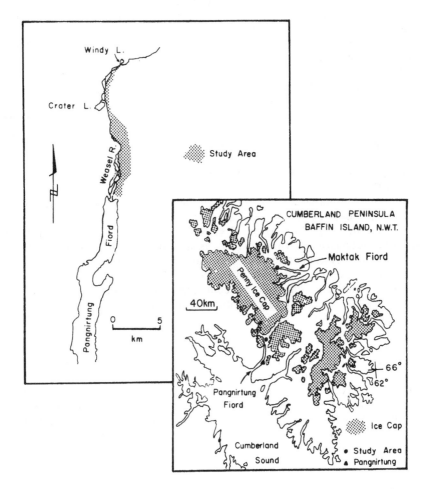

Figure 1 Location map showing the locations of study areas in Pangnirtung Pass and Maktak valley relative to the Penny Ice Cap and the Cumberland Peninsula.

Climate

The climate of southeastern Baffin Island is wholly arctic. Seasonal influences include the presence of sea ice, advection from the North Atlantic ocean, air mass incursions, and cyclonic and anticyclonic activity (Maxwell 1982). Locally, large differences in micro-climate exist as a result of variable exposure to wind and solar radiation, particularly within the deep fiords and valleys.

Winter is characterized by prolonged, severe cold (mean daily temperature -35°C) toward the north and in inland areas, while somewhat milder conditions prevail on the southeast coast (mean daily temperature -25°C). The atmospheric structure is generally characterized by strong thermal inversions such that upland areas are frequently warmer than the valleys below. Snowfall is associated with old Atlantic or mid-latitude cyclonic storms, although total precipitation is small.

A major change in circulation pattern takes place in June and is maintained until late September. The number and intensity of polar anticyclones drops dramatically with a corresponding increase in warm, humid cyclones from the west. Unlike the systems from the south, these bring thick cloud, and light and frequent precipitation. The presence of a cold, partially ice-covered sea and the low incidence of direct radiation results in cool summer temperatures with an average high of 10.5°C at Pangnirtung and an average low of 3.5°C.

In late September and through October, the relative frequency of cyclones from the south increases again until the winter pattern is re-established by November. Extensive cloud cover and light snow are characteristic of this period as a result of the vigour of these low pressure systems and the presence of open water.

On the average, Pangnirtung receives 34.8 cm (water equivalent) of precipitation each year as compared to an average of 25 cm/a for hot deserts. Snowfall accounts for about 54% of this amount. The average total receipt of snow is only 6.6 cm (water equivalent) for the period from December to March and mixing ratios are normally below 1.0 g/kg which illustrates the aridity of winters in the eastern Canadian Arctic. However, large interannual fluctuations are typical and mild winters may result from prolonged advection from the Atlantic Ocean. Approximately 17.5 cm (water equivalent) is received during the summer period (June to September) which is approximately three times that for the winter period. Mixing ratios are highest in July (about 18 g/kg) although very high moisture contents are never experienced.

Proglacial process controls

Moisture

Moisture as water, snow and ice is a key control of aeolian processes on the sandurs of Baffin Island, and clearly distinguishes this environment from those at more southern latitudes.

Water   Arctic rivers are dominated by rapid snow and ice melt during the short summer season.   Nival melt occurs in June and July, followed by a gradual decrease as less snow is available (Church 1974).   This pattern may be interrupted by peak flows associated with summer storms and direct surface runoff.   In watersheds where permanent icefields or snowpacks occur, peak runoff may be delayed until mid to late summer.   Such a regime is termed 'proglacial' by Church (1974).   Where a sandur occurs in these hydrologic regimes, a large proportion of its surface may be submerged during July and early August (Fig. 2), but exposed to wind action in late summer and autumn as the water levels fall and drying occurs through evaporation.

Figure 2 View of the completely submerged Weasel River sandur in early August 1985 near Crater Lake. All of the sediments in the foreground are aeolian. The arrow indicates the position of the time lapse camera.

Gravimetric moisture content analysis of fine to medium sand from the Weasel River sandur in Pangnirtung Pass measured in late August 1984 ranged from 19% to 24% (i.e. near saturation). McKenna-Neuman (1983) has shown that fine to coarse, well sorted sands are exceedingly resistant to deflation when pore water potentials exceed at most 5×10^5 N m^{-2}, or gravimetric moisture contents fall below approximately 1%. Although small dust devils do occur over the Weasel River sandur, they are isolated, infrequent and small. They seem to occur on days with clear skies and gusty winds when presumably strong surface heating and efficient water vapour transfer favours the evaporation of pore water to a depth of several grain diameters.

A similar analysis was performed on 180 surface samples from aeolian sand sheets and climbing dunes adjacent to the Weasel River sandur from mid July through August 1985. Of these samples,

75% had moisture contents less than 1.0% and 39% less than 0.1%. Although it would seem that these aeolian features are sufficiently dry to be reworked during the summer, measurements of the combined saltation and surface creep loads during the same period ranged from only 7.7 to 770.6 g/m-w. The low values reflect the light winds observed during this period (hourly values averaged 1.6 m/s at 10 m height and never exceeded 8 m/s).

Snow   Although winters on eastern Baffin Island are much drier than the summer period, snow cover on the sandur may also block or retard deflation.   The role of snow cover in shielding surface sediments from winter deflation was assessed with a time lapse camera located on an aeolian sand deposit near Crater Lake (Fig. 2).   One photograph was taken at noon each day on 16 mm film from August 25, 1984 to March 20, 1985.   The photograph from November 25, 1984 (Fig. 3) shows the surface of the aeolian sand deposit in the foregound and the Weasel River sandur in the background.   The bare surface of the deposit is typical of winter conditions for this year since snow cover was noted on only 26 out of 246 days of record.   The snow cover was never more than several centimeters deep and seldom covered the entire surface of the deposit.

Figure 3 Time lapse photograph of the Weasel River sandur and aeolian sand deposit on November 25, 1984. The painted poles which were positioned to record abrasion appear in the foreground.

Although Dietrich (1977) demonstrated that blowing ice crystals are agents of abrasion, preliminary observations in Pangnirtung Pass provide no evidence of this.   Poles covered with eight coats of exterior enamel were placed at various locations throughout the study area, including one site where abrasion could only occur by blowing snow and ice. Here the paint and its glossy appearance were perfectly preserved, while the paint and up to 1 mm of the hard maple dowel was stripped by the first week in December from poles placed on several of the aeolian deposits (Figs. 4A & B).

Figure 4 A Photograph of the abrasion of eight coats of exterior enamel from the north facing side of the pole which appears in Figure 3. The top of the aluminum base was originally level with the ground surface and is now deeply pitted. **B** The paint on the south facing side of the pole is perfectly preserved.

Ice During the period from October through May, pore water is frozen to cement the sand bodies. Because the surface of these deposits is snow free during most of the winter, some loss of ice by sublimation occurs slowly as dry air passes over the surface. The particles which are no longer strongly bonded by ice are much more susceptible to transport by wind and may eject other particles by impact as saltation occurs. Evidence for this mechanism is provided in the observation by Syvitski *et al.* (1983) of linguoid ripples on a frozen sandur surface which had been "sand-blasted flat".

Detailed field examination of the sublimation, deflation and abrasion of frozen surfaces is exceedingly difficult during the severely cold and dark winter conditions of the Canadian Arctic. Controlled laboratory simulations are needed to confirm the existence of these mechanisms and to expand understanding of them.

Ice cover on the braided river systems of arctic sandurs also varies seasonally, and therefore influences the aerial exposure of the sediments to wind action.   As indicated above, the Weasel River sandur is almost completely submerged in July and early August (Fig. 2).   In late autumn, the sandur obtains greater exposure as the water level decreases to its lowest point and flow stops (Fig. 3).   However, flooding of the sandur appears to have occurred during two periods of mild weather and strong winds in January.   During one 24 hour period on January 16 and 17, air temperature at Pangnirtung rose from -33°C to +1°C and the mean hourly wind speed increased from 0 to 16 m/s at 10 m height.   A similar event also occurred on January 7 and 8.   Figure 5, reproduced from the original colour, time lapse photograph, shows the icing that formed as a result of the flooding.   It covered most of the sandur and shielded the surface from further wind action.

Figure 5 Time lapse photograph for January 10, 1985 of the newly formed icing on the Weasel River sandur.

Arctic winds
While data indicate that arctic winds are generally not any stronger than those in temperate regions of Canada (Fig. 6), the strongest winds occur in sites where recordings have not been made.   In Pangnirtung Pass, local residents report very strong winds that limit or prevent travel, especially during the late autumn and spring (MacGregor 1924).   Two factors account for these winds.   Large but localized temperature gradients result in topographically generated wind systems including strong katabatic winds which are best developed over snow and ice surfaces.   Constrictions in air flow associated with the abrupt narrowing of the cliffed fiord walls or with large obstacles such as the Neoglacial moraines on the valley floor, create a venturi effect and locally very high wind

velocities (Oke 1978). Although no direct wind velocity measurements exist for Pangnirtung Pass, observations of extensive abrasion and ventifaction in discrete areas support the conclusion that the geomorphologically significant winds are very strong, but localized.

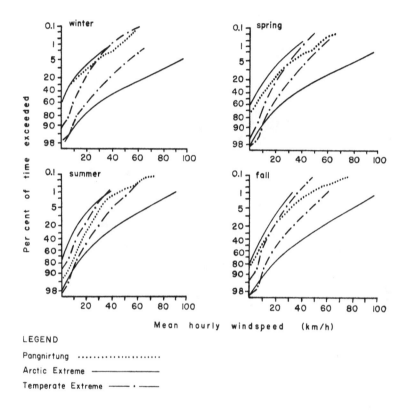

Figure 6 A comparision of extremes in windspeed frequency based on data from 38 arctic stations and 15 temperate stations across southern Canada. Aside from Resolution Island, the arctic extreme for strongest winds, regional winds in the arctic are not significantly stronger than those in temperate regions of Canada.

Aeolian landforms of the Pangnirtung and Maktak Valleys

Moisture controlled forms
Moisture controlled features in the eastern Arctic include niveo-aeolian deposits, adhesion structures and isolated ripples.

Snow banks buried by wind blown sand are commonplace at many of the fiord heads in eastern Baffin Island. The large aeolian deposits on the Pangnirtung Pass and Maktak valley floors are both at least partly snow cored (Fig. 7). While some deposits consist of alternating beds of sand and snow, others are mostly snow

Figure 7 Snow at the core of one of the largest aeolian sand sheets in south Pangnirtung Pass. The internal bed of aeolian sand adjacent to the scale marker indicates that these permanent snow banks are periodically buried by wind blown sand.

Figure 8 Hollow produced on the surface of the Maktak deposit as a result of the melting of snow underneath. Width of the hollow is 5 m.

containing thin, discontinuous lenses of sand and silt. The melting of the buried snow beds during warm summers produces hollows as shown in Figure 8. However, once the snow is incorporated in the permafrost, melting cannot occur.

The structures associated with pore water are comparatively small and ephemeral. Adhesion structures are occasionally formed in the summer months by windblown sand adhering to wet sandur surfaces. The coarse, discontinuous, sinuous ripples shown in Figure 9 are typical of features observed in other environments where there is a paucity of sediment for transport (Reineck & Singh 1980). The medium to fine sized particles in this particular surface are stabilized by pore water after a heavy rainfall the previous day. The coarse granules forming the ripple crests are rolled along by virtue   of their perched position on the surface, and the small magnitude of interparticle capillary forces relative to their weight.

Figure 9 Isolated, discontinuous, granule ripples forming over a wet, medium sand surface. Scale in lower right is 10 cm.

Forms controlled by surface roughness and regional topography
The majority of aeolian forms in Pangnirtung Pass are controlled by a continuum of surface roughness elements varying in size from a grain of sand (about 1 mm in diameter) to a boulder (about 2 m in diameter). The resultant forms include sand and granular ripples, sand tails and drifts, vegetation stabilized deposits, and lag-ventifact pavements. Many of these features are superimposed

upon larger aeolian deposits which include sand sheets and climbing dunes. The form and orientation of these larger deposits are clearly governed by the regional topography.

Aeolian sand sheets   Large, flat to gently graded aeolian sand deposits flank the sandurs in Pangnirtung Pass and at the head of Maktak Fiord. The largest is the Maktak deposit (Fig. 10) which extends approximately 4.9 km along the valley axis and has a maximum width of 900 m. The three Pangnirtung deposits are much smaller with a maximum length of 380 m.

Exposures from seven pits dug among the three Pangnirtung sand sheets indicate that the bedding generally consists of subcritically climbing translatent strata with tabular to wedge shaped units and planar bedding surfaces. The most distinctive sedimentary structures within the deposit are thin uniform bands of coarse sand, granules and even pebbles which appear to be the product of excessive deflation and granular ripple accretion (Fig. 11). These coarse bands are often more predominant towards the present surface and are bordered by fine to medium sands.

Figure 10 Aeolean deposit at the head of Maktak fiord. Photograph by M. Church.

Climbing dunes   Climbing dunes drape several of the glacial moraines in Pangnirtung Pass. Measurement of three of these features gives a range of 30 to 60 m for the basal width, and 12 to 30 m for the length of the inclined surface. The internal structure of these deposits predominantly consists of subcritical to

Figure 11 Pit exposure from the largest sand sheet in south Pangnirtung Pass showing coarse bands of granules interbedded with medium sands. Distance from top to bottom is 1.13 m.

supercritical climbing translatent strata with tabular to wedge shaped units and planar bedding surfaces, again indicating deposition through ripple migration. Ripple crest preservation is occasionally observed providing some evidence of grainfall deposition.

Sand and granular ripples Sand and granular ripples form much of the surface of the Pangnirtung sand sheets, although they are less common on the Maktak deposit. Granular ripples on the largest Pangnirtung sand sheet are cuspate in plan form and extremely asymmetric with the lee side facing south. The crests of these ripples appear as the bright areas in Figure 12. The largest ripples are approximately 9-13 cm in height and 3-4 m in length. Both of these dimensions increase from north to south.

Sand tails and sand drifts These leeside accumulations of windblown sand range in height and length from several millimeters (Fig. 13A) to several meters (Fig. 13B). Sand tails are particularly common near the peripheries of the climbing dunes and are entirely formed under winds from the south as are the climbing dunes themselves. In comparison, the sand drifts which accumulate to the lee of larger obstacles are exclusively formed under north winds. Often these obstacles are ventifacted with well polished flutes and

Figure 12 Granule ripples on the largest sand sheet in Pangnirtung Pass.

grooves on the windward side while the leeward face is typically lichen covered with no evidence of abrasion (Fig. 13B).

Aeolian deposits stabilized by vegetation
Vegetation stablized deposits are aerially more extensive in south Pangnirtung Pass than all other types of barren aeolian deposits combined.   They cover extensive areas of the valley sides in addition to the valley floor to which the barren aeolian deposits are limited.   The scale of these vegetated forms ranges from several centimeters for deposits trapped by individual plants to several meters for parabolic, dome and complex linear dunes (Figs 14A & B).   Arctic grasses, willows (*Salix* sp.) and the broad leafed willow herb (*Epilobium latifolium*) are the principal colonizers of the deposits on the valley floor while moss campion (*Silene acaulis*) and bilberry (*Vaccimium uligonosum*) are more common on the higher slopes.   Many of these plants are ideally suited to the trapping and accumulation of aeolian sand in that they are able to produce adventitious roots as the lower parts of their stems are buried.   However, abrasion can also severely damage a plant and discourage colonization as evidenced by the restriction of plant growth to the protected leeside of some drift deposits (Fig. 13B).

Lag-ventifact pavements
These features are exceedingly resistant to wind erosion as a result of a high concentration of immobile surface roughness elements

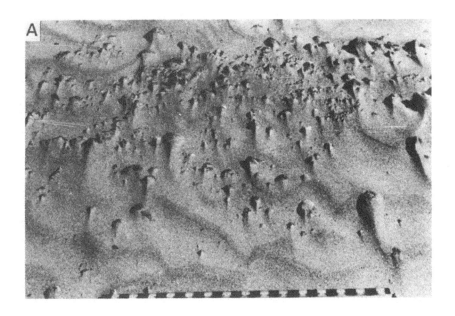

Figure 13 A Sand tails at the periphery of a climbing dune. Scale is in centimeters, **B** Sand drift in the lee of ventifacted boulder. The drift affords a suitable shelter for vegetation from the dessicating and abrasive effects of the wind.

Figure 14 A Wind blown sand trapped by a single willow (*Salix* sp.). The willow is growing in a down-wind direction toward the south. The north facing, windward side of the trunk and stems is badly damaged from abrasion. **B** A small parabolic dune with a well developed slip face and densely vegetated (*Salix* sp.) crest and arms.

which effectively shield other relatively small, mobile grains. They are the product of excessive deflation and are aerially extensive on the floor of south Pangnirtung Pass. Very mature ventifacts may or may not be present. The abrasion of only two of the eight layers of paint from poles erected on one of the cobble covered lag surfaces provides evidence of limited winter transport over these surfaces. Furthermore, the establishment of small sand patches and coppice features on very stable lag surfaces near the head of Pangnirtung Fiord suggests that there may be an upper limit to the stability of a lag surface beyond which the increasing roughness is conducive to the trapping and accumulation of sand particles. The increasing stability of the lag surface also appears to favour plant growth which in turn traps even more incoming sand.

Grain size

Grain size statistics based on the method of moments were evaluated for a variety of depositional environments within the Pangnirtung and Maktak study areas. Mean, standard deviation and skewness are plotted in Figures 15 and 16. Kurtosis was also calculated, but a preliminary investigation indicated that this statistic was not environmentally sensitive.

Grain size statistics for ripple crests, ripple troughs and planar surfaces on the Pangnirtung and Maktak sand sheets are plotted in Figure 15. The most environmentally sensitive factor is mean grain size varying from 0.10 to 4.35 mm for ripple crests, 0.28 to 2.71 mm for ripple troughs and 0.68 to 2.42 mm for planar surfaces. Mean grain size is generally coarser and clearly more variable for the ripple crests than for either the ripple troughs or the planar surface. Many of the trough samples are negatively skewed which is likely the result of larger granules rolling down into the troughs from the coarse ripple crests. Negative skewness is uncommon for either the ripple crest or planar surface samples. The sand sheet samples are typically well sorted to moderately sorted with standard deviation values between 0.25 and 0.82 mm. The slightly shorter 'needles' extending from points for the ripple trough data in Figure 15B indicate marginally better sorting than for either the ripple crests or the planar surfaces.

Grain size statistics for the climbing dunes, sand drifts and vegetated aeolian deposits are compared to those for the planar surface samples from the sand sheets in Figure 16. Mean grain size is again the most environmentally sensitive factor. The mean grain size of the vegetated deposits (Fig. 16A) ranges from 0.01 to 0.64 mm which is considerably finer than the range for the sand sheets (Fig. 16B). However, samples from vegetation trapped sand on the valley floor is clearly coarser than that from the valley side with a mean grain size of 0.30 mm as compared to 0.15 mm for the latter. Sand collected from the leeside of vegetated dunes is

A RIPPLE CRESTS

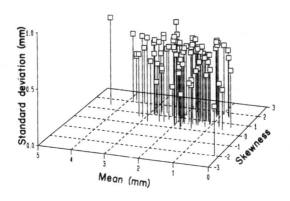

B RIPPLE TROUGHS

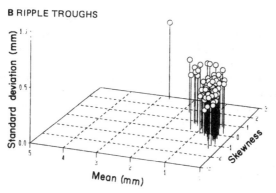

C SAND SHEET PLANES

Diamond Planar surface, Pangirtung
Heart   Planar surface, Maktak

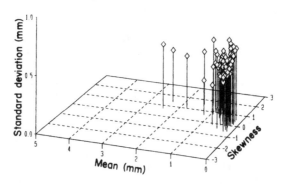

Figure 15 Plots of grain size statistics (mean, standard deviation and skewness) for ripple crests A, ripple troughs B, and planar surfaces C of the Pangnirtung and Maktak sand sheets.

A VEGETATED AEOLIAN DEPOSITS

Club   Vegetated Dune, leeside   Square Vegetation trapped sand, valley floor
Diamond Vegetated Dune, windward side   Balloon Vegetation trapped sand, valley side
Cross   Vegetation trapped sand, Maktak   Star   Vegetated dune, crest

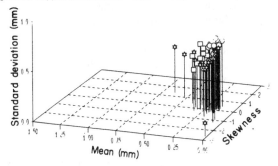

B SAND SHEETS PLANES

Diamond Planar surface, Pangnirtung
Heart   Planar surface, Maktak

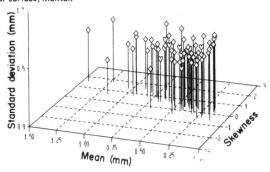

C CLIMBING DUNES AND SAND DRIFTS

Square   Climbing dune, transect upslope   Flag   Sand drift deposits
Balloon   Climbing Dune, E-W Transect   Cross   Inter-drift surface

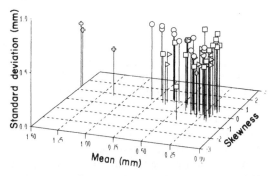

Figure 16 Plots of grain size statistics (mean, standard deviation and skewness) for vegetated aeolian deposits **A**, planar sand sheet surfaces at the appropriate scale **B** provided for comparison and climbing dunes and sand drifts **C**.

comparable in mean grain size to that from the valley side features, whereas that from the dune crests is variable. Mean grain size for samples collected from the climbing dunes and sand drifts in the Pangnirtung study area (Fig. 16C) is comparable in size and spread to that for the planar surfaces of all the sand sheets. Samples from inter-drift surfaces are generally coarser than either the dunes or the drifts. The planar surfaces, dunes, drifts and vegetated deposits cannot be distinguished on the basis of either skewness or standard deviation, although the samples appear to become slightly more positively skewed with increasing mean grain size.

Conclusions

(a) Although they are uncommon in the uplands, aeolian landforms are prevalent on the valley floors of glaciofluvial outwash areas in eastern Baffin Island.

(b) Aeolian transport on arctic proglacial sandurs is potentially most significant during the fall and early winter months when both the aerial exposure of the sediments and wind velocities are greatest.

(c) Ministry of Transport weather data recorded at Pangnirtung indicate that two periods of above freezing temperatures and strong winds occurred in January 1984. Time lapse photographs of an area of the Weasel River sandur provide evidence that these above average temperatures were coincident with a flooding of the sandur. The resultant icing shielded the sandur from further wind action.

(d) The supply of sediment for winter transport on Baffin Island sandurs may be provided by two mechanisms: (i) the grain-by-grain release of particles from the frozen sandur surface as a result of the sublimation of interstitial ice and (ii) the abrasive action of grains saltating across the frozen surface. Theoretical consideration and laboratory simulations are needed to examine the nature and efficacy of these mechanisms.

(e) Although Dietrich (1977) demonstrated that blowing snow crystals are agents of abrasion, preliminary observations in Pangnirtung Pass provide no evidence of this.

(f) The aeolian deposits flanking the Weasel River sandur in south Pangnirtung Pass include sand sheets, climbing dunes, sand drifts and vegetated accumulations. Unlike the deflation of the sandur which is strongly controlled by seasonal variations in moisture, these deposits are relatively dry throughout the summer and bare of snow during the winter, and therefore, are reworked to some extent year round. Topography, surface roughness and vegetation strongly control the form of these deposits, and as a result, they resemble many of the forms observed in warmer environments.

(g) The majority of aeolian landforms in south Pangnirtung Pass and Maktak valley appear to be formed exclusively by winds travelling down valley toward the fiord heads. Since these directions are respectively north and ESE, it is suggested that (i) the effect of topography on local wind patterns overprints regional wind patterns and (ii) the generation of the most geomorphically significant winds is likely related to the relative proximity of the Penny ice Cap and its outlet glaciers. Year round observations of field equipment in south Pangnirtung Pass indicate that abrasion was also limited to the north facing side and to fall and early winter activity. This suggests that although reworking of the aeolian deposits may occur year round, the most active period is the same as for the sandur.

(h) With the exception of the vegetated features on the valleyside, mean grain size measurements from aeolian deposits in south Pangnirtung Pass and the Maktak valley are generally substantially coarser than the 0.125 to 0.25 mm range typical of temperate inland dunes. The very fine sand and silt fraction of the glaciofluvial source sediments is poorly represented in these deposits. Field measurements are required to determine whether or not this fine fraction is borne in suspension much further downwind, and deposited on the frozen surface of the fiord during winter.

Acknowledgements

This work was supported by a grant from the Natural Sciences and Engineering Research Council of Canada. We gratefully acknowledge the help of Parks Canada, especially Mr. Ray Breneman, chief warden of Auyuittuq National Park. J. Gaskey, V. Horvath and P. Neuman provided invaluable assistance in the field and laboratory. Environment Canada provided the anemometer and the time lapse camera.

References

Adie, R.J. (ed.) 1970. *Antarctic geology and geophysics.* International Union of Geological Sciences, Series B, Number 1.

Andrews, J.T. and A.S. Dyke 1974. *A geomorphological inventory and interpretation of Baffin Island National Park and surrounding area.* Unpublished Final Report to Applied Research Division, Parks Canada, Contract 73-66.

Andrews, J.T., P.J. Weber, and H. Nichols 1979. A late Holocene pollen diagram from Pangnirtung Pass, Baffin Island, N.W.T.,

Canada. *Review of Paleobotany and Palynology* 27, 1-28.

Church, M. 1970. *Baffin Island sandur: a study of arctic fluvial environments.* Unpublished Ph.D. Thesis, University of British Columbia.

Church, M. 1974. Hydrology and permafrost with reference to northern North America. *Permafrost Hydrology, Proceedings of Workshop Seminar* 7-20. Ottawa: Canadian National Committee of the International Hydrological Decade.

Davis, P.T. 1985. Neoglacial moraines on Baffin Island. In *Quaternary environments eastern Canadian Arctic, Baffin Bay and western Greenland*, J.T. Andrews (ed.), 682-718. Boston: Allen and Unwin.

Dietrich, R.V. 1977. Impact abrasion of harder by softer materials. *Journal of Geology* 85, 242-6.

Dyke, A.S. 1979. Glacial and sea-level history, southwestern Cumberland Peninsula, Baffin Island, Canada. *Arctic and Alpine Research* 11, 179-202.

Embleton, C. and C.A.M. King 1975. *Periglacial geomorphology.* London: Arnold.

Gilbert, R. 1982. Contemporary sedimentary environments on Baffin Island, N.W.T., Canada: glaciomarine processes in fiords of eastern Cumberland Peninsula. *Arctic and Alpine Research* 14, 1-12.

Gilbert, R. 1983. Sedimentary processes of Canadian arctic fiords. *Sedimentary Geology* 36, 147-75.

Gilbert, R. 1984. The fiordlands of northeastern Canada. In *Studies in Canadian regional geography. Essays in honor of J. Lewis Robinson*, B.M. Barr (ed.), 27-37. B.C. Geographical Series Number 37. Vancouver: Tantalus Research Ltd.

MacGregor, W.B. 1924. *Annual report of the Royal Canadian Mounted Policy*, 1924.

Maxwell, J.B. 1982. *The climate of the Canadian arctic islands and adjacent waters.* Volume 2. Environment Canada, Atmospheric Environment Service.

McKenna-Neuman, C. 1983. *A combined theoretical and wind tunnel investigation of the effect of capillary water upon the threshold of movement of sand grains by wind.* Unpublished M.Sc. Thesis, University of Guelph.

Mullen, R.E., D.A. Darby and D.L. Clark 1972. Significance of atmospheric dust and ice rafting for Arctic Ocean sediment. *Geological Society of America Bulletin* 83, 205-12.

Nickling, W.G. 1978. Eolian sediment transport during dust storms: Slims River Valley, Yukon Territory. *Canadian Journal of Earth Sciences* 1069-84.

Oke, T.R. 1978. *Boundary layer climates.* London: Methuen.

Reineck, H.-E. and I. B. Singh 1980. *Depositional sedimentary environments.* Berlin: Springer-Verlag.

Selby, M.J., R.B. Rains and P.W. Palmer 1973. Eolian deposits of the ice-free Victoria Valley, Southern Victoria Land, Antarctica. *New Zealand Journal of Geology and Geophysics* 17, 543-62.

Shaw, J., R. Gilbert and J.J. Archer 1978. Proglacial lacustrine sedimentation during winter. *Arctic and Alpine Research* 10, 689-99.

Syvitski, J.P.M., K.W. Asprey, C.P. Blakeney, D. Clattenburg and G.D. Hodge 1983. SAFE: 1982 Delta Report. *Sedimentology of arctic fjords experiment.* Compiled by J.P.M. Syvitski and C.P. Blakeney. HU 82-031 data report. Volume 1 Canada Data Report, Hydrography and ocean Sciences, No. 12, 18-1 to 18-41.

13

The frequency and source areas of dust storms

Nicholas John Middleton and Andrew S. Goudie
and Gordon L. Wells

Abstract

Dust events are important geomorphological phenomena in many of the world's dry lands and four main types can be defined: dust storms, dust haze, blowing dust and dust devils. This paper is concerned with the first two of these categories. Methods for their investigation include the examination of terrestrially observed meteorological data, remote sensing, and deep-sea core sediments. Meteorological data are widely available, but care needs to be exercised in their use. Three main types of remote sensing platforms have proved to be useful: geostationary meteorological satellites, polar orbiting satellites, and the space shuttle orbiters. The deep-sea record provides a long sequence of deposition and helps in the identification of major sinks. Combining the data from these three sources, the major locations of dust storm activity can be identified. However, dust storm frequencies are prone to variation from year to year, and land-use and climatic changes can substantially affect their occurrence. Among important controls of the distributional pattern are the meteorological conditions that entrain dust, including low pressure fronts with intense baroclinal gradients, the convergence zone between cold air masses in regions of monsoonal airflow, katabatic winds, and local convectional cells (haboobs). Another major control is the nature of the surface upon which deflational activity works.

Introduction

The importance of dust storms in the world's dry lands has become increasingly apparent in recent years, and with this appreciation has developed a clearer picture of the great significance of deflation and deposition in desert and peri-desert environments. Among the reasons for this development of interest (see Goudie

1978, 1983, Péwé 1981, Morales 1979a, Coudé-Gaussen 1982) are the increasing availability of meteorological data in the era of air transport, the study of dust storms from space, and the realization that dust storms have many environmental implications for such phenomena as soil erosion, climatic change, loess formation, and stream loads (Goudie 1983, Table 1). The purpose of this paper is to examine certain key aspects of dust storms: their definition, methods for their investigation, their frequencies both now and in the past, the meteorological situations that produce them, and the environments from which deflation occurs with relative facility. In view of their geomorphological significance, it is important to appreciate some of the problems associated with establishing the criteria by which their frequency, magnitude and areal extent can be quantified.

Definition

In many parts of the world, visibility may be reduced for a variety of reasons, including fog, smog, smoke, as well as aeolian entrainment of dust. The degree of visibility reduction can cover a range from light haze with visibility of many kilometers, to severe events with a visibility of just a few meters. Thus, in defining dust storms it is necessary to isolate the role of dust in relation to other possible mechanisms of visibility reduction, and to impose some visibility limit. The following terms may be appropriate in defining dust events created by aeolian processes:

(a) *Dust storms* are the result of turbulent wind systems entraining particles of dust into the air, so that visibility is reduced to less than 1000 m. Examination of humidity data may differentiate visibility reduction resulting from wind from that resulting from fog.

(b) *Dust haze* consists of aeolian dust particles suspended in the air, which are not actively being entrained and which have been raised from the ground by a dust storm that occurred prior to the time of observation or at a considerable distance away. Visibility may sometimes be reduced to less than 1000 m, but is frequently more than that figure.

(c) *Blowing dust* is the state where dust is raised above the ground locally, but does not reduce visbility at eye level to less than 1000 m.

(d) *Dust devils* or dust whirls are local, spatially limited, columns of dust that neither travel far nor last long.

In this paper we are primarily concerned with the first two of these categories.

Table 1 Major global dust source areas with key station dust storm day frequency (visibility < 1,000m a year (D)).

Source Area	Reference to figure 4	Station	D	Reference	Data Period	Number of Years
Australia						
Central Australia	A	Alice Springs	15.6	Aus. Bureau of Met.	1942-82	41
China						
Takla Makan	B	Hetian	32.9	Chinese Central Met. Bureau	1953-80	28
Kansu Corridor	C	Minqin	37.3	"	1953-80	28
USSR						
Turkmenia	D	Repetek	65.5	Klimenko & Moskaleva (1979)	1936-60	25
Kara Kum	E	Nebit Dag	60.0	"	"	"
Rostov	F	Zevetnoe	23.3	"	"	"
Altai	G	Rubtsovsk	25.1	"	"	"
Alma Alta	H	Bakanas	47.7	"	"	"
Kazakhstan	I	Dzhambeiti	45.9	"	"	"
SW Asia						
Thar desert	J	Fort Abbas (Pakistan)	17.8	Middleton 1986a	1951-58	8
Upper Indus plains	K	Jhelum (Pakistan)	18.9	"	1951-58	8
Afghan Turkestan plains	L	Chardarrah	46.7	"	1974-80	7
Seistan Basin	M	Zabol (Iran)	80.7	"	1967-73	7
Makran coast	N	Jask (Iran)	27.3	"	1970-73	4
Middle East						
Lower Mesopotamia	O	Kuwait Int. Airport	27.0	Safar (1985)	1962-84	23
N. Saudi/Jordan/Syria	P	Abou Kamal (Syria)	14.9	Middleton (1986b)	1959-79	21
N. Africa						
Bodele Depression	Q	Maidurguri (Nigeria)	22.5	Nigerian Met. Dept	1955-79	25
S. Mauritania/N. Mali/C. Algeria	R	Nouakchott (Mauritania)	27.4	Service Met. Mauritania	1960-84	25
Libya and Egypt	S	Sirte (Libya)	17.8	Libyan Met. Dept.	1956-77	22

Methods of dust storm investigation

Study of individual dust events and long term dust storm frequencies involves three main methods of investigation that may be used independently or in conjunction: terrestrially observed meteorological data, data from remote sensing platforms and reconstructions from deep sea-cores.

Successful tracking of individual events has been carried out in a few studies using the WMO network of SYNOP stations (e.g. Bertrand 1977, Morales 1979b). SYNOP codes delimit a range of dust events by visibility, whether dust is raised locally or advected from afar, and some element of event scale, as well as other simultaneous meteorological readings. The only compatible world wide data base for dust storm activity over an extended period, however, is in the form of dust storm days, (defined as a day in which a dust storm is recorded) which in the large majority of countries involves a visibility reduction to below 1000 m. However, the use of such data involves a number of problems that should be noted.

A dust storm day does not indicate the duration of dust events at a resolution finer than the 24 hour period, and gives no indication of the areal or volumetric extent of raised dust. Any observed data will involve some element of subjective judgement by the observer (e.g. differentiating dust from smoke), and a "dust storm day" may blur the important distinction between locally raised dust (dust storm sensu stricto) and that blown from elsewhere (dust haze sensu stricto). The availability of data may present problems, particularly in remote areas where station densities are low and where observation frequencies during each day are limited. Over large areas of arid and semi-arid terrain, stations may be completely absent. Indeed, at stations in some countries where dust storms occur, such events may not be recorded at all (e.g. in Chile, Peru, South Africa), and in others, data may be unavailable for essentially political reasons (e.g. Mongolia). Nevertheless, meteorologically observed dust storm days provide a useful indication of the world wide distribution of dust storm activity, as shown below, and form a valid data base from which to launch more detailed local investigations.

Satellite tracking, problems, space shuttle photography
Various kinds of remote sensing imagery obtained from Earth orbit offer the best opportunity for calculating the source regions, areal extents and trajectories of major dust events. Useful images are collected by three types of orbiting platforms: geostationary meteorological satellites (e.g. GOES, Meteosat, GMS), polar orbiting satellites (e.g. NOAA-n series, DMSP series) and the Space Shuttle orbiters. Each type of system has its merits and weaknesses when the data are used to attempt to monitor dust outbreaks.

Meteorological satellites in geostationary orbit keep a constant watch over continental-scale regions of the Earth. Though their data transmission is continuous, the spatial resolution is typically quite low (4 to 20 kilometers), and their visible and thermal radiometer bandwidths are broad. The thermal instruments are designed to detect the extent of relatively cold cloud tops contrasting against warm ocean and land surfaces. As a result, warm tropospheric dust palls often create insufficient temperature contrast to be detected by the thermal radiometers on board meteorological satellites (Fig. 1). The low spatial resolution of the visible data from these platforms makes dust identification difficult. In many cases, the disappearance of high-contrast terrain features usually detected by these sensors is the signal to a dust event, rather than an obvious well-defined dust plume seen on the imagery. Only the largest and densest dust systems are well-documented by geostationary satellites (e.g. Noyalet 1978).

The Advanced Very High Resolution Radiometer (AVHRR) on board the NOAA Polar Orbiter series and similar instruments on board the DMSP series image large regions from low Earth orbit. Though the AVHRR is in continuous operation and can be monitored by ground stations within its footprint, only limited recorder time is available to collect data from remote parts of the world for retransmission to ground stations in the U.S.A. This lack of recording capacity coupled with the sun-synchronous polar orbits prohibits constant monitoring of regions with frequent dust activity, limiting image collection to one or two orbital passes per day. The AVHRR may be operated with 1 or 4 kilometers spatial resolution and has visible, near-infrared and thermal radiometers with medium bandwidths. Though the presence of dust is not always unambiguous in the thermal channels, the visible and near-infrared radiometers permit precise determination of dust extent and trajectory during daylight passes (Fig. 2). Given its 2200 kilometers swath width and approximately 2000 kilometers scene length, the Polar Orbiter AVHRR is capable of providing detailed information about pan-regional dust events.

The most impressive orbital imagery comes as the result of the training of STS astronauts to recognize and to photograph dust events with hand-held cameras. Their visible colour film images may achieve 20 meters spatial resolution in nadir views, and can provide stereoscopically-overlapping frames for three-dimensional analysis. During STS missions, surface station reports and meteorological satellite data are monitored for dust occurrences. Messages are transmitted to the astronaut crews to alert them of orbital passes over dust events. Conversely, the astronauts often report and photograph dust events not recorded by other data sources. While STS missions offer only periodic regional coverage due to launch schedules, orbital parameters and conflicting crew activities, astronaut photography has provided crucial evidence of dust dynamics, and a means for evaluating the success of other

Figure 1 Meteosat-2 thermal infrared image (top) gives no indication of the degree of atmospheric turbidity over the eastern Mediterranean and Israel on 17 April 1985 at 0600 GMT. An STS astronaut photograph (bottom) made at 06:16–04 GMT on the same day shows the extent of an African dust pall not seen in the thermal radiometer data.

data sources for providing information concerning the occurrence, scale and trajectory of dust events.

One of the most important contributions of the astronaut photography has been to examine the accuracy of dust reports from surface meteorological stations. In remote regions of areas such as western Africa, blowing sand and dust may be reported by fewer than one-quarter of the surface stations in the dust pall area photographed by astronauts. Similarly, the astronaut photographs,

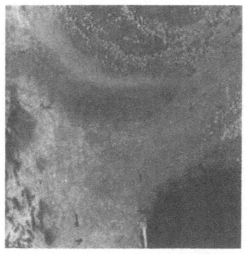

Figure 2 A NOAA-7 Advanced Very High Resolution Radiometer (AVHRR) image shows a dust pall swept by a cyclonic gyre from the Southern High Plains across central Texas on 4 April 1984.

with the time of exposure recorded on the frames by the data modules, can be used to evaluate simultaneous visible and thermal radiometer data gathered by geostationary meteorological satellites (Fig. 1). The photographs may eventually lead to new means for calculating the mass budget of dust events. From surface dust collection, Yaalon and Ganor (1979) have estimated dust densities of 10 tonnes km^{-3} for a single dust storm while Junge (1979) has argued that the global annual contribution of dust to atmospheric aerosols may range from 130 to 800 million tonnes per year. By examining the astronaut dust photographs exposed at a known time with a given sun elevation and viewing geometry, it may be possible to derive a measurement of relative optical depth in relation to clear areas of the frame, especially when the photograph has been taken over areas with a stable reference surface of known characteristics, such as the ocean. Such a measurement would yield the minimum dust density required to completely obscure surface features. If such a measurement can be obtained, the imagery from orbital remote sensing platforms will be capable of providing information on dust event trajectory, areal extent and density.

The deep-sea core record

Because the deep-sea core record provides a long sequence of deposition, it has proved possible to determine areas of major aeolian dust input by investigating ocean floor sediments. Sarnthein and Koopman (1980), by analyzing the grain-size distribution of the coarser than six micron fraction (carbonate free) off West Africa, have been able to demarcate areas of maximum aeolian inputs both for the Holocene and for the late glacial maximum (ca 18 000 years BP). They suggest that at 18 000 years BP, there was a huge area of aeolian silt deposited up to 800 km offshore from West Africa, with sedimentation rates approximately double those of today. High rates of silt accumulation are also evident in ocean sediments of late glacial age off Australia (Thiede 1979) and the Arabian and Thar deserts (Kolla & Biscaye 1977). There are, however, other areas where aeolian inputs are much less, including the Atlantic offshore from the Namib (Diester-Haass 1985). Rates of aeolian sedimentation in the Pacific are also relatively low, but nonetheless indicate long continued contributions from Asian deserts, with a crescendo of aeolian activity in the Pleistocene (Duce *et al.* 1980, Leinen & Heath 1981, Rea *et al.* 1985). It is now possible to identify dust source areas as far back as the Cretaceous, and for much of that period the Sahara appears to have been a dominant source area (Lever & McCave 1983). As Figure 3 shows, annual mass budgets can be calculated for Transatlantic dust events derived from the Sahara using rates of sediment accumulation and aerosol collection from ships.

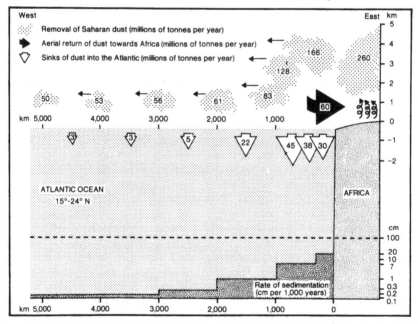

Figure 3 Aeolian sediment budgets for the Sahara (modified after the work of Coudé-Caussen and others)

Global distribution of major dust source areas

There have been a number of attempts to delimit the major world source areas of dust production (e.g. Grigoryev & Kondratyev 1980, Péwé 1981, Goudie 1983, Coude-Gaussen 1984). In Figure 4, we attempt to develop these world maps further to subdivide the world areas into more limited dust sources, and quantify them by identifying those that experience more than 15 dust storm days a year. This map is complemented by Table 1 which names the major sources and shows the mean number of dust storm days at stations that represent the scale of activities in these regions, using the longest available meteorological records.

The major world dust producing regions are located in the broad band of arid and semi-arid land stretching from West Africa to northern China. Probably the most important dust sources in the world are located in the Sahara Desert, which can be subdivided into five main regions: the Bodélé Depression alluvial plains in Niger and Chad, source of the Harmattan dust wind; an area that comprises southern Mauritania, northern Mali and central southern Algeria; southern Morocco, and western Algeria; the southern fringes of the Mediterranean Sea in Libya and Egypt; and northern Sudan.

In the Middle East, two major dust sources can be identified: the alluvial plains of southern Mesopotamia, source of the dust in the Shamal wind, and the desert areas of Syria, Jordan and northern Saudi Arabia. In Iran, the Makran coast is a major dust source, as is the Seistan Basin on the borders with Afghanistan and northwest Pakistan. The Thar Desert and the upper Indus plains are major sources in northwestern India and in Pakistan. In central Asia, an important source is located on the loess and alluvial plains of the Amu Darya in northern Afghanistan. In Soviet central Asia, major sources are in the Kara Kum Desert and on the shores of the Caspian Sea in Turkmenia. North of this area are the plains of Kazakhstan, to the west the Rostov region, between the Caspian and Black Seas, and to the east the regions of Alma Alta and Altai. In northern China, four dust sources are identifiable: The Takla Makan Desert in the Tarim Basin; the Kansu Corridor; the loess plateau of Inner Mongolia; and the Gobi Desert, crossing the border into southern Mongolia.

Less important world sources are located in the central deserts of Australia, the southwest USA, and northern Mexico. In South America, the Argentine pampas is an important source, but data are not available to assess the dust output from the Atacama Desert. Similarly, in southwestern Africa data are unavailable, but lack of widespread appropriate dust producing geomorphological units suggests that it is not a dust producer on the scale of the North African sources. The Somali peninsula may be an important source, but is also unquantifiable due to a lack of data.

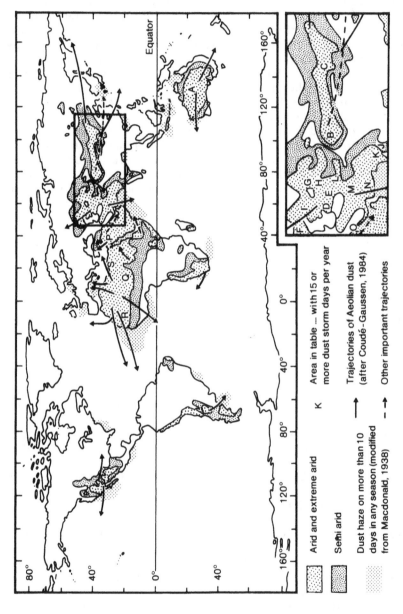

Figure 4 The global distribution of intensive dust storm activities, and major dust trajectories.

Arid and extreme arid

Semi arid

Dust haze on more than 10 days in any season (modified from Macdonald, 1938)

K   Area in table .... with 15 or more dust storm days per year

→   Trajectories of Aeolian dust (after Coudé - Gaussen, 1984)

--→   Other important trajectories

Changes in dust storm frequency

Although the major world areas of dust-raising activity have been outlined above, it is important to note that the frequency and intensity of dust output from these regions may vary quite considerably from one year to the next, quite apart from seasonal variations.   On a geological timescale this variation may be identified in the ocean sediment record (see above) or the polar ice caps (Thompson 1977).   Since the beginning of routine and standardized meteorological observations, shorter term variations, on a time scale of decades, become apparent.   The spatial extent of such variations range from the stabilization or destabilization of a local lake bed for example, to a region the size of the Sahelian fringe of the southern Sahara.   Some examples are illustrated in Figure 5.

The great increase in dust-raising activity during the 1930s in the Great Plains of the USA is well documented.   In this case, overuse and mismanagement of the soil by pioneer agriculturalists was triggered by a number of drought years to produce a dramatic increase in dust storm occurrence, as shown in Figure 5A for Dodge City, Kansas (Chepil *et al.* 1963).   During the Virgin Lands scheme in the USSR in the 1950s, a similar increase in dust storms was experienced as land was stripped of vegetation and ploughed. Table 2 shows this increase in activity for five stations.   This sort of destabilization of potentially wind-erodible surfaces by agricultural development has been reported from many parts of the world: on the loessic soils of the Chaco region of northern Argentina (Prego 1961); in a number of Californian valleys (Clements *et al.* 1963); and currently on the semi-arid steppes of northern Mongolia (Sanders personal communication 1982).

Table 2 The effects of the Virgin Lands scheme on frequency of dust storm days in the Omsk region of the USSR (after Sapozhnikova 1973).

Station	Mean annual number of dust storm days		
	1936-1950	1951-1962	Increase
Omsk, steppe	7.0	16.0	x 2.3
Isil'-Kul'	8.0	15.0	x 1.9
Pokrov-Irtyshsk	4.0	22.0	x 5.5
Poltavka	9.0	12.0	x 1.3
Cherlak	6.0	19.0	x 3.2
mean	6.8	16.8	x 2.5

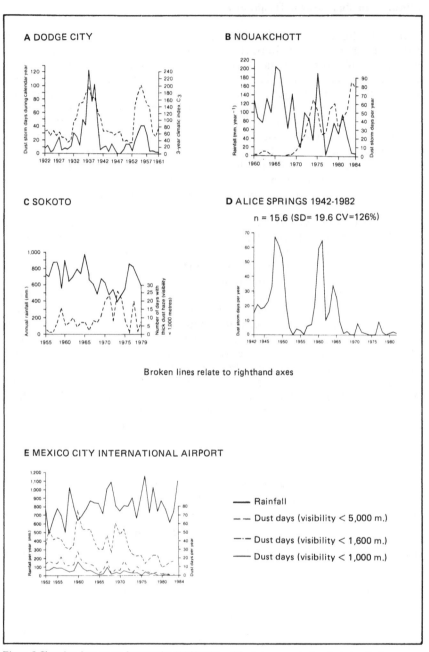

Figure 5 Changing dust storm frequencies (A) Number of dust storms and climatic index C_3 for Dodge City, Kansas (after Chepil *et al.* 1963). (B) Annual dust storm frequencies (dotted line) and annual rainfall (solid line) for Nouakchott, Mauritania (data from Service Meteorologique, Nouakchott). (C) Annual dust haze frequencies for Sokoto, Nigeria (D) Annual dust storm day frequency for Alice Springs, Australia (E) Annual dust days at Mexico City International Airport.

In many cases, human activities are just one factor affecting variations in dust storm frequencies. The run of years with below average rainfall in the Sahel, that started in the late 1960s and continued into the 1980s, has been instrumental in producing a marked increase in dust storms from this vast region (Middleton 1985). It seems likely, however, that the effects of the current drought have been exacerbated by the actions of man as an agent of desertification - through overgrazing, overcultivation, range burning and the chopping of wood for cultivation and fuel. This situation is illustrated at Nouakchott, Mauritania, in Figure 5B where declining rainfall that started in 1968 marks the initiation of a prolonged increasing trend in the number of dust storm days. At Sokoto, in northern Nigeria, a definite upward phase in the number of days with thick Harmattan dust haze dating from 1966, peaking in 1971 and 1973 and dropping again by 1976, is mirrored by a downward phase in rainfall over the same period (Fig. 5C).

Droughts in other parts of the world may result in an increase in dust storm activity over several years. At Alice Springs, central Australia, dry years in the late 1940s and early 1960s had a dramatic effect on the number of dust storm days (Fig. 5D).

At a more localized scale, man may act to reduce dust storm frequency. Mexico City was plagued by intense dust storms blown from the dry bed of Lake Texcoco that was drained as the urban area was developed. In the 1950s, this dust source accounted for 40% of dust storms over the city, but due to the combined effects of a gradual increase in rainfall dating from the 1940s, an expansion of the urban area around the lake bed and a project specifically set up to stabilize the lake bed in 1972 by revegetation

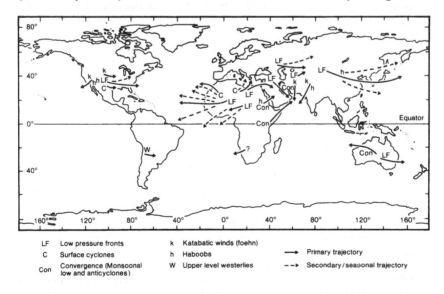

Figure 6 Global map of the main synoptic meteorological conditions associated with dust events.

and the construction of reservoirs, the frequency of dust storms over the city has been severely curtailed (Fig. 5E), and by 1984 dust storms from Lake Texcoco were completely eliminated.

Synoptic meteorological conditions

Dust-raising events may occur under a wide variety of meteorological conditions within most global regions experiencing dust storms. However, the most frequent and severe dust storms typically occur under only one of a few synoptic meteorological conditions prevailing over any selected region. A number of dust-generating weather systems have been identified. By far the most important is the passage of low pressure fronts with intense baroclinal gradients that are accompanied by very high velocity winds entraining and carrying dust. Surface cyclones themselves may sweep out gyres of dust, if circulation around the low pressure is sufficiently intense (Fig. 6). In regions of monsoonal airflow, dust may be funnelled along the convergence zone between cold air masses associated with regions of low pressure and hot, tropical anticyclonic air masses. More localized dust storms occur when katabatic winds deflate alluvial plains and fans adjacent to regions with considerable topographic relief. In at least one region, the high base level altitude permits the upper westerly airflow to create point-source dust storms at the surface. Finally, the classic Haboob creates local dust storms by the cold downburst winds generated by large convective cells.

In Australia, dust storms generally follow the passage of strong low pressure fronts tracking eastward across the southeastern portion of the continent (Loewe 1943). During the summer monsoon, the convergence zone between high and low pressure systems may serve to channel dust from the interior of the Simpson Desert across Alice Springs and out over the Indian Ocean. Such a convergence may occur simultaneously with the movement of a low pressure front across southeastern Australia (Sprigg 1982). In addition, tropical cyclones which cross the northern coastline of Australia may generate dust as they track inland towards the dry interior (as with Hurricane Cecile in March 1984).

From western China across eastern Eurasia, low pressure fronts transport dust aerosols over vast areas, and material may be carried in the upper westerlies to the Alaskan arctic (Rahn et al. 1981) and into the Pacific Basin (Ing 1972). Local storms produced by katabatic winds may occur in the Tarim and Quaidam Basins, while upper level westerly airflow likely generates point-source dust storms over the +4000 m Tibetan Plateau. Haboobs are known to occur in the Gobi Desert and are probably generated in the Kansus region.

Moving across Eurasia to the arid Soviet steppes and interior deserts, low pressure fronts following an easterly trajectory are again the primary agents of long-distance transport. Katabatic airflow may be locally important such as the garmsil wind that blows down the northern facing slopes of Kopetdag, raising dust in Turkmenia (Nalivkin 1982). Crossing the Hindu Kush and Karakoram Ranges, katabatic winds deflate the plains of the Indus and its tributaries, and the Quaternary lakebeds and alluvial fans of Afghanistan and eastern Iran. These point-source dust storms can be characterized by extremely high velocity surface winds and dense palls (Middleton, 1986a). Within the Thar Desert of India and Pakistan, dust is transported by the westerly loo wind in spring, the result of a strong pressure gradient brought about by a deepening of the seasonal trough, and haboobs, known locally as "andhi" (Joseph et al. 1980).

The major meteorological conditions promoting dust storms in the Middle East are depressions moving eastwards from the Mediterranean across Turkey, the Levant and northern Iraq. The shamal winds lifting dust from Iran and adjacent regions are usually associated with low pressure anchored over southern Iran that forms a strong baroclinal gradient with a semi-permanent anticyclone over northern Saudi Arabia. The convergence zone between the two pressure systems induces high velocity, turbulent winds for regional dust transport during a time of intense convection over the Tigris-Euphrates floodplain due to very high surface temperatures (Membery 1983).

The convergence associated with the summer monsoon over eastern Africa channels dust from Ethiopia, Somalia and northern Kenya across the Arabian Sea to the region north of Karachi as seen on NOAA Polar Orbiter AVHRR imagery. Farther north, in Sudan, similar convergence creates dust storms crossing the Red Sea into Saudi Arabia, while the classic haboob of the Khartoum area remains the primary dust-moving system (Freeman 1952). In Morocco, Algeria and Libya, intense depressions may sweep bands of dust across the eastern Atlantic and central Mediterranean. Low pressure fronts tracking across the north African coast carry dense dust palls to the Middle East (Yaalon & Ganor 1979). In southern Africa, dust movement in Namibia tends to be caused by the "berg winds", a warm off-plateau, partially katabatic phenomenon.

By far the most important global dust transport occurs with the passage of low pressure fronts across the southern Sahara and Sahel. The harmattan winds may carry half of all the world's mineral aerosols (Junge 1979). These depressions tend to track along a southerly course during the northern hemisphere winter with more zonal easterly transport occurring during summer months. Throughout the year, trans-Atlantic export of dust from the Sahara may occur (Prospero & Nees 1977, Prospero et al. 1981).

In the mountainous regions of western North America, local katabatic winds generate dust storms in California (Santa Ana winds, Bowden et al. 1974) and along the Colorado Rocky Mountain Front. Haboobs are the primary dust-raising meteorological event in Arizona (Nickling & Brazel 1984), and may occasionally occur in the southern High Plains. In the famous Dust Bowl region of the High Plains, low pressure fronts tracking eastward may transport aeolian materials from agricultural regions of Texas and New Mexico into the Atlantic Ocean off the southeastern coast (Henz & Woiceshyn 1980, McCauley *et al.* 1981). Surface cyclones crossing Texas may also raise dust palls (Fig. 2).

On the Andean Altiplano of Chile, northwestern Argentina and southern Bolivia, dust is raised from salars and alluvial fans by superimposed westerlies across the region, with localized deflation accentuated by airflow around peaks (Fig. 7). Dust is raised in the valleys of the Argentine foothills by katabatic airflows known as "zonda", and on the loessic plains of the pampas by eastward moving frontal passage (Wolcken 1951).

Environments of deflation

The susceptibility of surfaces to deflation varies greatly, but very few sound empirical data are available to ascertain the critical threshold velocities for the input of soil particles into the air. A major attempt to rectify this problem was made by Gillette *et al.* (1980) employing a specially developed portable wind tunnel which permits the estimation of minimum threshold velocities under field conditions. They found that the velocities increase with different types of soil in the following sequence: disturbed soils, sand dunes, alluvial and aeolian sand deposits, disturbed playa soils, skirts of playas, playa centers, desert pavements.

However, given that the prime component of dust storms is silt-sized material dominated by quartz, it is necessary to have some mechanisms whereby such silty materials can be generated. Among these mechanisms are weathering of bedrock and sand grains by salt processes, simulations of which have been undertaken successfully in the laboratory (Goudie *et al.* 1979, Pye & Sperling 1983, Goudie in press). There is also evidence that silt-sized material is produced by salt weathering in the neighbourhood of saline playas (Goudie & Day 1981). However, there are other mechanisms for silt production in arid areas including glacial grinding in places like the Karakorams, frost shattering, spalling during fluvial transport, abrasion during wind transport (Whalley *et al.* 1982) and chemical weathering along structural weaknesses in quartz (Nahon & Trompette 1982).

Given that there are so many mechanisms to produce silt-sized material, it is not surprising that various geomorphological environments contain silt-sized material that is available for deflation. These include situations like outwash and alluvial fans, playa basins, weathered or unconsolidated exposures and areas of previously deposited loess (Fig. 8).

Coudé-Gaussen (1984), whose work is largely based on the Sahara, has attempted to categorise desert surfaces that are highly

Figure 7 Many point-source dust storms are raised by westerly airflow across the playa surfaces and alluvial aprons of the Andean altiplano in northwestern Argentina on 30 August 1983. In the original colour photograph, the plumes originating from pan evaporites are white, while those from alluvial fans are tan. (NASA/Astronaut Hasselblad).

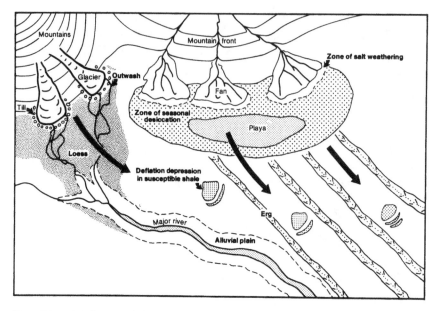

Figure 8 A model of geomorphological environments from which substantial deflation occurs.

favourable for producing dust:

    (a) Les cuvettes de sebkhas - dried out salt lakes of internal drainage, the surface of which is disrupted and rendered mobile by salt crystallization;

    (b) Les epandages d'oueds - wadi sediments containing silt and the floodplains of great rivers like the Niger;

    (c) Les surfaces de fech-fech - powdery areas derived from ancient lake muds or on certain argillaceous rocks;

    (d) Les takyrs - desert clay soils with polyglonal desiccation cracks;

    (e) Les affleurements rocheux meubles - outcrops of rocks like unconsolidated Neogene fine-grained sediments.

The importance of such environments can be assessed by looking at the locations of meteorological stations that show high levels of dust storm activity in relation to their geomorphological setting, and by investigating the patterns of blowing dust revealed on satellite imagery. Large numbers of pertinent photographs are currently being provided by the STS astronauts. Among major source areas are the following, some of which have been confirmed by the distinctive mineralogy of their dusts:

    (a) The alluvial plains of the Tigris-Euphrates system (Khalaf & Al-Hashash 1983).

(b) Lake sediments of the Bodélé Depression in the Sahara (McTainsh & Walker 1982), and fans from Tibesti and Hoggar.

(c) Alluvial plains of the Niger River and recently devegetated ancient dunes of southern Mauritania.

(d) The alluvial and loessic deposits of the Upper Indus plains.

(e) The salars and associated fans of the Andean region (Fig. 9).

(f) The ancient coversands and fluvial sediments of the High Plains of the USA.

(g) Playa erosion in the Great Basin of the USA (Young & Evans, 1986).

(h) The closed basins and fans of the Seistan Basin (Middleton, 1986a).

(i) The alluvial and lacustrine deposits of the Aral-Caspian system (Kes 1983).

(j) The great loess belt of China.

(k) The fans, dune fields and playas of the Tarim basin.

(l) The Lake Eyre basin and its feeding alluvial plains.

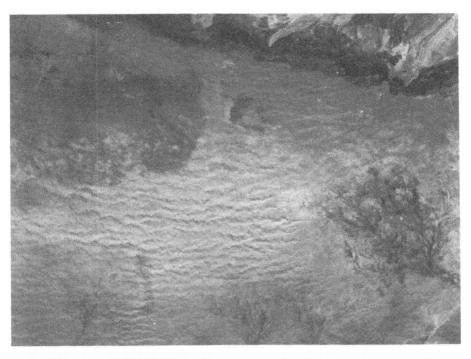

Figure 9 A detail from a Large Format Camera photograph made during Space Shuttle mission 41G on October 1984 reveals the source of dust plumes in the alluvial plain of the Amu Darya River in northern Afghanistan.

Conclusion

Aeolian deflation, as represented by dust events, is a broad scale process for which it is necessary to derive some quantitative means for discovering frequency, magnitude and areal extent. Some success at obtaining an indication of regional frequencies and favoured environments for deflation has been achieved by the analysis of standard meteorological data, while analysis of ocean cores has provided long term data on those areas which are major offshore sinks for dust. Orbital reconnaissance techniques allow specific dust sources to be pin-pointed, provide a check on the reliability and comprehensiveness of meteorological data, but above all, give a true appreciation of the scale of dust transport. Nonetheless, further development of methodology is required to provide more reliable and comprehensive estimates of the volume, density and mass budgets of dust events at a global scale, and for this a combination of remote sensing, and ground and sea-borne collection is required.

References

Bertrand, J. J. 1977. *Action des Poussieres Sub-Sahariennes sur le pouvoir Glacogene de l'air en Afrique de l'ouest.* Thesis, Univeristy of Clermont-Ferrand, France.

Bowden, L.W., J.R. Huning, C.F. Hutchinson and C.W. Johnson 1974. Satellite photograph presents first comprehensive view of local wind: The Santa Ana. *Science* 184, 1077-8.

Chepil, W.S., F.H. Siddoway and D.V. Armbrust 1963. Climatic index of wind erosion conditions in the Great Plains. *Soil Science Society of America, Proceedings* 449-51.

Clements, T., R.O. Stone, J.F. Mann and J.L. Eymann 1963. *A study of windbourne sand and dust in desert areas.* U.S. Army Natik Laboratories, Earth Sciences Division, Technical Report ES-8.

Coudé-Gaussen, G. 1982. Les poussieres eoliennes sahariennes, mise au point. *Revue de Geomorphologie Dynamique* 21, 49-69.

Coudé-Gaussen, G. 1984. Le cycle des poussieres eoliennes desertiques actuelles et la sedimentation des loess peridesertiques quaternaires. *Bulletin Centre Recherches et Exploration-Production Elf-Aquitaine* 8, 167-82.

Diester-Hass, L. 1985. Late Quaternary sedimentation on the eastern Walvis Ridge, SE Atlantic (HPC 532 and four piston cores). *Marine Geology* 65, 148-89.

Duce, R.A., C.K. Unni, E.J. Ray, J.M. Prospero and J.T. Merrill 1980. Long range atmospheric transport of soil dust from Asia to the tropical North Pacific: temporal variability. *Science* 209, 1522-4.

Freeman, M.H. 1952. Duststorms of the Anglo-Egyptian Sudan. *Meteorological Reports* No. 11. H.M.S.O. London.

Gillette, D.A., J. Adams, A. Endo and D. Smith 1980. Threshold velocities for input of soil particles into the air by desert soils. *Journal of Geophysical Research* 85, 5621-30.

Goudie, A.S. 1978. Dust storms and their geomorphological implications. *Journal of Arid Environments* 1, 291-310.

Goudie, A.S. 1983. Dust storms in space and time. *Progress in Physical Geography* 7, 502-30.

Goudie, A.S. In press. Laboratory simulation of "the wick effect" in salt weathering of rock. *Earth Surface Processes and Landforms.*

Goudie, A.S. and M.J. Day 1981. Disintegration of fan sediments in Death Valley, California by salt weathering. *Physical Geography* 1, 126-37.

Goudie, A.S., R.U. Cooke and J.C. Doornkamp 1979. The formation of silt from quartz dune sand by salt weathering processes in deserts. *Journal of Arid Environments* 2, 105-12.

Grigoryev, A.A. and K.J. Kondratyev 1980. Atmospheric dust observed from space. *World Meteorology Organization Bulletin* 30, 3-9.

Henz, J.F. and P.M. Woiceshyn 1980. *Climatological relationships of severe dust storms in the Great Plains to synoptic weather patterns.* Jet Propulsion Laboratory Publication 79-97. Californian Institute of Technology.

Ing, G.K.T. 1972. A dust storm over central China, April 1969. *Weather* 27, 136-45.

Joseph, P.V., D.K. Raipal and S.N. Deka 1980. "Andhi", the convective duststorm of northwest India. *Mausam* 31, 431-42.

Junge, C. 1979. The importance of mineral dust as an atmospheric constituent. In *Saharan dust*, C. Morales (ed.), 49-60. Chichester: Wiley.

Kes, A.S. 1983. Study of deflation processes and transfer of salts and dust. *Problemyi Osvoeniya Pustin* 1, 3-15.

Khalaf, F. and M. Al-Hashash 1983. Aeolian sedimentation in the northwestern part of the Arabian Gulf. *Journal of Arid Environments* 6, 319-32.

Klimenko, L.V. and L.A. Moskaleva 1979. Frequency of ocurrence of dust storms in the USSR. *Meteorologiya i Gidrologiya* 9, 93-7.

Kolla, V. and P.E. Biscaye, 1977. Distribution and origin of quartz in the sediments of the Indian Ocean. *Journal of Sedimentary Petrology* 47, 642-9.

Leinen, H. and G.R. Heath 1981. Sedimentary indicators of atmospheric activity in the northern hemisphere during the Cenozoic. *Palaeogeography, Palaeoclimatology, Palaeoecology* 36, 1-12.

Lever, A. and I.N. McCave 1983. Eolian components in Cretaceous and Tertiary North Atlantic sediments. *Journal of Sedimentary Petrology* 53, 811-32.

Loewe, F. 1943. Duststorms in Australia. *Australian Meteorological Bureau Bulletin* No. 28.

McCauley, J.F., C.S. Breed, M.J. Grolier and D.J. MacKinnon 1981. The U.S. dust storm of February 1977. In *Desert dust: origins, characteristics and effects on man*, T.L. Pewe (ed.) 123-47. Geological Society of America, Special Paper 186.

McTainsh, G.H. and P.H. Walker 1982. Nature and distribution of Harmattan dust. *Zeitschrift fur Geomorphologie* 26, 417-35.

Membery, D.A. 1983. Low level wind profiles during the Gulf Shamal. *Weather* 38, 18-24.

Middleton, N.J. 1985. Effect of drought on dust production in the Sahel. *Nature* 316, 431-34.

Middleton, N.J. 1986a. A geography of dust storms in southwest Asia. *Journal of Climatology* 6, 183-96.

Middleton, N.J. 1986b. Dust storms in the Middle East. *Journal of Arid Environments* 10, 83-96.

Morales, C. (ed.) 1979a. *Saharan Dust*. Chichester: John Wiley.

Morales, C. 1979b. A review of weather systems connected with dust storms in the Sudan and surrounding areas. *Rapporter och notiser* 41.

Nahon, D. and R. Trompette 1982. Origin of siltstones: glacial grinding versus weathering. *Sedimentology* 29, 25-35.

Nalivkin, D.V. 1982. *Hurricanes, storms and tornadoes*. New Delhi: Amerind.

Nickling, W.G. and A.J. Brazel 1984. Temporal and spatial characteristics of Arizona dust storms (1965-1980). *Journal of Climatology* 4, 645-60.

Noyalet, A. 1978. Utilisation des images mcteosat: genese et evolution d'une tempete de sable sur l'ouest Africain. *La Meteorologie* 14, 113-5.

Péwé, T.L. (ed.) 1981. *Desert dust: origin, characteristics, and effect on man*. Geological Society of American Special Paper 186.

Prego, A.J. 1961. La erosion eolica en la Republica Argentina. *Ciencia e Investigacion* 17, 307-24.

Prospero, J.M and R.T. Nees 1977. Dust concentration in the atmosphere of the equatorial North Atlantic: possible relationship to the Sahelian drought. *Science* 196, 1196-8.

Prospero, J.M., R.A. Glaccum and R.T. Nees 1981. Atmospheric transport of soil dust from Africa to South America. *Nature* 289, 570-2.

Pye, K. and C.H.B. Sperling 1983. Experimental investigation of silt formation by static breakage processes: the effect of temperature, moisture and salt on quartz dune sand and granitic regolith. *Sedimentology* 30, 49-62.

Rahn, K.A., R.D. Borys and G.E. Shaw 1981. Asian desert dust over Alaska: anatomy of an arctic haze episode. In *Desert dust: origins, characteristics and effects on man*, T.L. Pewe (ed.), 37-70. Geological Society of America, Special Paper 186.

Rea, D.K., M. Leinen and T.R. Janecek 1985. Geologic approach to the long-term history of atmospheric circulation. *Science* 227, 721-5.

Safar, M.I. 1985. *Dust and duststorms in Kuwait.* Kuwait: Directorate General of Civil Aviation.

Sapozhnikova, S.A. 1973. Map diagram of the number of days with dust storms in the hot zone of the USSR and adjacent territories. FSTC-HT-23-0027-74. U.S. Army Foreign Science and Technology Center, Charlottesville, Virginia.

Sarnthein, M. and B. Koopmann 1980. Late Quaternary deep-sea record on northwest African dust supply and wind circulation. *Palaeoecology of Africa* 12, 239-53.

Sprigg, R.C. 1982. Some stratigraphic consequences of fluctuating Quaternary sea level and related wind regimes in southern and central Australia. In *Quaternary dust mantles of China, New Zealand and Australia*, R.J. Wasson (ed.) 211-40. Canberra: Australian National University.

Thiede, J. 1979. Wind regimes over the late Quaternary southwest Pacific Ocean. *Geology* 7, 259-62.

Thompson, L.G. 1977. *Microparticles, ice sheets and climate.* Institute of Polar Studies Report 64, Ohio State University.

Whalley, W.B., G.R. Douglas and J.P. McGreevy 1982. Crack propagation and associated weathering in igneous rock. *Zeitschrift fur Geomorphologie* 26(1), 33-53.

Wolcken, K. 1951. Descripcion de una violenta tempestad de polvo. *Meteoros* 1, 211-16.

Yaalon, D.H. and E. Ganor 1979. East Mediterranean trajectories of dust-carrying storms from the Sahara and Sinai. In *Saharan dust*, C. Morales (ed.), 187-93. Chichester: Wiley.

Young, J.A. and R.A. Evans 1986. Erosion and deposition of fine sediments from playas. *Journal of Arid Environments* 10,103-15.

14

Effect of antecedent moisture conditions on dust storm generation in Arizona

Anthony J. Brazel, William G. Nickling and Jeffrey Lee

Abstract

Meteorological weather data are used to identify historical dust events for southern Arizona, U.S.A. to classify synoptic conditions, and to characterize the moisture environment over the period 1948-1984.

Dust frequencies vary spatially in southern Arizona primarily in response to surface moisture-related factors, although links to synoptic controls have not been definitively analyzed. The role of short-term (less than a season in length) antecedent moisture variability through time does not appear to play a major role in dust generation. However, climatologists have not yet been able to fully resolve meso- to local scale precipitation patterns on a scale necessary to adequately portray a spatial antecedence pattern for dust storm processes. A longer-term, regional moisture antecedence effect is detectable from the climatic data and suggests that winter moisture is inversely related to annual dust frequencies at Phoenix, Arizona. However, several processes which are not possible to capture from available meteorological data (e.g. anthropogenic disturbance effects) probably control the large unexplained variance in the temporal pattern of dust frequencies.

Introduction

Dust storms are frequent in the American Southwest, but vary considerably in magnitude and in their seasonal occurrence within the Sonoran Desert region of the U.S.A. At a national scale, dust storm climatologies have been produced by Orgill and Sehmel (1976) and Changery (1983), and only a few detailed dust analyses have been developed for the Southwest Desert region such as those of

Ingram (1976), Péwé (1981), Nickling and Brazel (1984), and Brazel and Nickling (in press).

The purpose of this paper is to review the dust climatology of the Arizona desert environment and to explore the role of antecedent factors in the production of dust events.

Dust events in Arizona

Tables 1 and 2 summarize characteristics of Arizona dust events based on three first order weather stations in southern Arizona, those of Phoenix, Tucson, and Yuma. A dust event is recorded by the National Weather Service when visibility is reduced by dust to less than 11.3 km. Dust storms, on the other hand, are defined as

Table 1 Arizona dust storms by month.*

	YUMA		PHOENIX		TUCSON	
Month	% per Month	% Time Blowing Dust	% per Month	% Time Blowing Dust	% per Month	% Time Blowing Dust
J	3	.12	0	.00	0	.00
F	3	.11	1	.02	0	.00
M	6	.18	1	.01	0	.00
A	17	.28	3	.13	0	.00
M	6	.17	6	.03	9	.01
J	3	.03	8	.05	37	.01
J	18	08	28	.15	27	.02
A	27	.17	35	.20	27	.01
S	12	.21	15	.15	0	.00
O	0	.00	2	.00	0	.00
N	2	.03	1	.01	0	.00
D	3	.10	0	.00	0	.00

* Relative to less than 1.6 km visibility obstruction, as defined in Nickling and Brazel (1984)

events when visibility is reduced to less than 1.0 km. These data are useful in determining the frequency of dust events in a region. However, the data do not allow the researcher to determine the exact sources of blowing dust, although insights can be gained by looking at the prevailing weather and obtaining knowledge on upwind land surface conditions. The papers of Brazel and Nickling (in press) and Nickling and Brazel (1984) have utilized an extensive dust event data base developed by Changery (1983). An analysis of first order weather stations for the southern, drier half of Arizona has been undertaken. Table 1 illustrates for Yuma, Phoenix, and Tucson two characteristics of dust frequencies. The first column under each station lists percent dust events distributed monthly at

Table 2 Characteristics of Arizona dust storms (1965-1980).

Type	YUMA				PHOENIX				TUCSON			
	Time (h)	Direction (degrees)	Time (min)	Duration (min)	Time (h)	Direction (degrees)	Time (min)	Duration (min)	Time (h)	Direction (degrees)	Time (min)	Duration (min)
Intense dust storms ⩽ 1.6 km visibility												
1A	1325* (410)†	250 (142)	140 (117)	208 (172)	1429	260	42	78	–	–	–	–
1B	1620 (378)	291 (20)	299 (190)	428 (254)	1534 (342)	204 (91)	132 (102)	229 (220)	–	–	–	–
2	1439 (748)	140 (87)	9 (14)	75 (63)	1834 (431)	129 (79)	9 (19)	54 (37)	1813 (144)	106 (46)	4 (4)	17 (17)
3	1419	110	16	258	1847	150	26	75	1715	110	14	13
4	1459	240	125	218	1708	238	23	32	1530	110	9	12
Moderate to weak dust storms > 1.6 km, ⩽ 11.3 km visibility												
1A	1427 (512)	283 (110)	–	138 (102)	1646	340	–	54	–	–	–	–
1B	1228 (675)	283 (50)	–	248 (197)	1507 (672)	260 (35)	–	135 (82)	118	233	–	261
2	1122 (927)	167 (70)	–	106 (70)	1637 (703)	167 (87)	–	57 (72)	–	–	–	–
3	2140	260	–	95	–	–	–	–	–	–	–	–
4	1619 (782)	270 (56)	–	214 (84)	1631	227	–	103	–	–	–	–

* mean
† standard deviation where appropriate
1A Pre-frontal dust
1B Post-frontal dust
2 Thunderstorm/Convective
3 Tropical Disturbance
4 Upper level/Cut-off low

each of the stations. Obvious in the data results is the effect of strong mesoscale convective activity in the summer months, typical of the so-called monsoon season of the American Southwest (Ingram 1976). Yuma, located toward the heart of the Sonoran Desert, displays a secondary peak in spring. These frequencies, however, should more appropriately be converted to percent time of blowing dust per month, thus incorporating the effect of duration as well as frequency. In Table 1, the second column under each station listing gives the percent time in minutes that dust was observed for each month. Note the sharp increase in peak frequency at Yuma, a secondary peak at Phoenix, and the overall relative insignificance of blowing dust at Tucson. Table 2 provides more detail on the synoptic types associated with dust events, the time that events are frequently occurring, the prevailing direction of blowing dust, the rate in minutes of visibility reduction from 11.3 km to the minimum observed during the event, and the overall storm duration in minutes.

These data provide useful insights into dust generation mechanisms and the role of meteorology in promoting blowing dust (Nickling & Brazel 1984). However, the data do not provide answers to some fundamental questions. For example, meteorological factors which are necessary to dust generation are, however, not always sufficient for dust generation. In other words, the high winds associated with virtually all synoptic situations identified may not necessarily obtain required threshold speeds to entrain dust, due to large scale armoring of the surface or dense vegetative cover acting to suppress dust generation.

The role of synoptic climatology in dust generation

Several case studies of mesoscale convective systems promoting individual dust events have been studied for Arizona (e.g. Idso *et al.* 1972, Ingram 1976, Hales 1975). However, no analysis other than that of Brazel and Nickling (in press) has revealed an overall insight into whether there is a significant climatological relationship between the occurrence of synoptic systems and blowing dust frequencies. Brazel and Nickling (in press) have produced a synoptic data set to gain insights into the relationship. Table 3 gives a summary of the role of synoptic climatology in the generation of dust. Synoptic weather types were determined by procedures used after Muller (1977). In Table 3, Dust/Total is the ratio between the number of dust events which are associated with a partĭcular synoptic type and the total frequency of that synoptic type for the period 1965-1980. If the forcing of the synoptics were the most important factor determining variance of dust frequencies among the sites, the Dust/Total ratios should be relatively constant from one location to the other. Clearly we see a ranking of this ratio for Tucson, Phoenix, and Yuma which relates to the annual precipitation ranking among the sites. The variation of the ratio among synoptic types at a site is more complex and is likely a function not only of the timing of these events in relation to seasonal moisture, but also the internal variability of wind and weather conditions that take place within a given synoptic type (Wigner & Peterson 1985). Further analyses on this aspect need to be undertaken to disentangle the complexities inherent in the synoptic systems themselves.

Table 3 Ratios of dust events by weather type (Dust) to total annual occurrence of that weather type (Total), 1965-1980.

Weather Type	YUMA			PHOENIX			TUCSON		
	Dust	Total	Dust/Total	Dust	Total	Dust/Total	Dust	Total	Dust/Total
1A & B	103	404	0.25	17	463	0.04	4	432	0.01
2	44	80	0.55	85	304	0.28	7	544	0.01
3	14	176	0.08	8	176	0.05	2	176	0.01
4	3	16	0.19	2	16	0.13	2	16	0.13
All Types	164	676	0.24	112	959	0.12	15	1168	0.01
Annual Precip (mm)		67			180			278	

1A & B Frontal types
2 Thunderstorm/Convective type
3 Tropical Disturbance Type
4 Upper level/Cut-off low

The role of precipitation

The work of many researchers has illustrated that dust frequencies relate to moisture characteristics of a given region (Goudie 1978, 1983, Pollard 1978, Middleton 1984). Our analysis includes relationships between dust events and antecedent precipitation. Phoenix, Arizona is chosen for the analysis.

Phoenix (population approximately 1.8 million in the metropolitan area) is located in the Salt River Valley, one of the many basins in the Basin and Range Physiographic Province of the U.S.A. Fault-block mountains in the region range from 300 to 1000 m above the basin and are made up of Precambrian metamorphic and intrusive rocks and Tertiary volcanic rocks. Sediments in the basin are mainly silt and clay, and silt is an abundant matrix in the colluvial-alluvial sediments on the flanks of the valley (Péwé 1981). Ephemeral streams are widespread, but major perennial rivers are dammed and rarely contain water in their lower reaches. Natural vegetation in the region is composed of Southwest desert shrubs of the lower Sonoran life zone (Lowe 1964). A large irrigated agricultural area is centered in the Phoenix region and includes

1855 square kilometers of the basin floor. This is only one percent of the sparsely vegetated desert region of southern Arizona.

Analysis of the temporal variation of dust frequencies includes:

(a) a climatology of the number of consecutive dry days less than a stated daily precipitation value antecedent to dust events;
(b) a determination of the spatial precipitation resolution in the development of the antecedent precipitation climatology in the Phoenix region; and
(c) the longer term seasonal and annual antecedent effects on Phoenix dust event frequencies.

Short-term antecedent dryness at Phoenix, Arizona

Table 4 presents for the Phoenix weather service office (which is located at an international airport that observes dust as an obstruction) a characterization of antecedent dryness for six daily precipitation values. The body of the table represents percent frequency. The upper left corner of the table indicates that prior to dust events at Phoenix, 83% of the time there was rain of at least 1.27 mm less than 50 days prior to dust events. In the lower right of the table, we see that 52% of the time there was rain of at least 25.4 mm ranging from 300 to 350 days prior to dust events.

Table 4 Percent frequency of antecedent numbers of consecutive dry days with precipitation less than stated value at Phoenix.

Daily Precip (mm)	Dust Antecedent Day Interval							
	0-50	50-100	100-150	150-200	200-250	250-300	300-350	350
1.27	83	13	4					
2.54	74	17	6	3				
6.35	51	16	16	10	5	2		
12.70	29	14	16	13	7	8	13	
19.05	17	10	8	11	11	12	31	
25.40	8	4	2	4	8	1	52	11

* Date extracted from continuous records at Phoenix WSO Airport station data (1948-1978). Number of dust events is 479.

Although increasing frequencies occur in longer antecedent time categories the larger the precipitation values become, we do not gain a sense of the significance of these antecedence data on dust generation. Dry days prior to dust events are typically substantial in number and probably exceed the time required to evaporate precipitated amounts of moisture derived from previous storms prior to blowing dust events. For example, in summer, approximately 6 to

7 mm per day evaporation can occur from clay, loam, bare soil (Idso *et al.* 1979).

A measure of the significance of these antecedent conditions on blowing dust is the comparison of antecedent conditions prior to dust events with antecedent conditions prior to non-dust days chosen at random in the summer months. Seventy days were chosen at random and 70 dust event days were chosen at random during the period 1948-1980 (a 15% sample of dust days recorded). A calculation of antecedent dryness was performed on these two data sets. A non-parametric Kolmogorov-Smirnov test, using a Chi-Square test statistic at a .05 level of significance, was used to determine if the two antecedent dryness data sets were significantly different. The data sets were composed of a 1 to 30 day antecedence data set and a 30 to 180 day antecedence data set. Although this test is not conclusive, results indicate there does not appear to be a significant difference at both these time scales in the timing of antecedent precipitation for non-dust days and dust days.

The problem of rainfall distribution

Part of the problem of detecting the role of antecedent rainfall on short time scales (up to seasons in length) relates to the Phoenix Airport station point estimate of regional rainfall, particularly since most dust events at Phoenix occur in the summer convective season. An analysis of correlations between Phoenix Airport rainfall and rainfall received at all surrounding stations was produced by Brazel and Ziriax (1979), using the methodology of Sharon (1974). The average distance decay of the correlation coefficient calculated between all stations and the Phoenix Airport indicates an insignificant correlation beyond approximately a 25 km radius of the Phoenix Airport station.

We took all dust events since 1948 and determined the number of consecutive antecedent days prior to each event back in time to precipitation days of 1.27, 2.54, 6.35, 12.70, 19.05, and 25.40 mm. We did this for both the Phoenix Airport station and 33 surrounding stations within a 70 km radius of Phoenix Airport. A series of correlations were produced on how far back in time would be required to attain a precipitation day of 1.27 mm to 25.40 mm between Phoenix Airport and the mean of all surrounding sites. The results indicate that there is a correlation decay ranging from a correlation coefficient of 0.62 for days of little rain prior to dust events, to a correlation coefficient of 0.15 for days of high rainfall prior to dust events. Thus, the Phoenix Airport station attains a significant correlation with the region as a whole for the number of antecedent days back to small daily precipitation totals. However, there is an insignificant correlation between Phoenix and its surroundings for the number of antecedent days (relative to dust events) back in time to very high rainfall amounts (days which may be more significant to suppressing large scale blowing dust).

Furthermore, we have selected 10 dust events where significant measureable rainfall occurred only a day or two before at the Phoenix Airport station. A determination was made of antecedent rainfall along a defined dust storm swath into Phoenix. We looked at a storm swath 70 km in length and one quadrant in azimuth upwind from the Airport station. Consecutive antecedent days with less than 2.54 and 6.35 mm were calculated for both the Phoenix Airport and for the storm track region. The consecutive number of antecedent dry days, on average, is substantially larger for the storm track region compared to the number of antecedent dry days determined for the Airport station.

Longer term regional antecedence

The dust data set has too small a sample size to look for more than one season's antecedent effects. In the usual manner, we have correlated annual precipitation with annual dust frequency (Goudie 1978). However, based on data from 1948-1984, a correlation coefficient of only 0.09 was obtained, not significant at the 95% confidence level. In addition to rainfall we have obtained regional Palmer Drought Severity Index (PDSI) data from Karl and Knight (1985) for climatic divisions of Arizona on monthly and seasonal time scales. Divisions are regions of relative homogeneity of climate in Arizona. The PDSI is a measure of ground moisture (Mather 1974, Palmer 1965). A correlation analysis was performed between antecedent seasonal PDSI values of winter, spring, summer, and fall and seasonal dust frequencies. There is a statistically significant correlation between previous winter regional PDSI in Central Arizona and total dust storms for a year, observed in Phoenix. The correlation coefficient value is 0.37, significant at the 95% confidence level, with n = 37.

We view such a relationship as reflecting the probable influences of moisture on several surficial processes, such as surface crusting (Gillette *et al.* 1982) and interannual variability of vegetative cover (Martin & Turner 1977, Karpiscak 1980). These processes would substantially affect threshold wind velocities and thus the frequency of large scale blowing dust in the region. Vegetative cover variability differences between extended dry and wet periods in Central Arizona has recently been shown to approximate 30 to 35% in river valleys and on bajada and alluvial fan slopes around the Phoenix region (Miller in press).

Summary

In the Southwestern U.S.A. only recently have dust climatologies been assembled. Accompanying synoptic information indicates that the forcing of wind action is, of course, necessary for dust generation. However, the surficial environmental conditions are highly variable spatially and temporally and these conditions also determine dust generation. Antecedent dryness prior to dust events which have occurred in Central Arizona is not well characterized by single station data where dust observations are made, but must be viewed along the dust storm track. High evaporation rates in summer tend to offset short term, moderate wet periods that may be antecedent to dust events. A longer term antecedence of moisture appears to control dust storm frequency in Central Arizona. Discernible at a moderate level of significance is a relationship between winter moisture and annual dust frequency. Evidence of large scale vegetative response to increases of moisture on these time scales appears to suggest that interannual variability of vegetation cover partially controls dust generation.

Conclusions

Blowing dust is a complex function of a variety of interacting environmental factors related to climate, soil conditions, and vegetative parameters. In addition, anthropogenic effects exacerbate the blowing dust problem. In the heavily populated region of Central Arizona, all of these factors interact to determine the frequency patterns of blowing dust. The determination of the contribution of each factor to the incidences of blowing dust large enough in scale to be observed at a standard weather station still remains a research goal. Field analysis of suspected source regions (e.g. river channels, disturbed sites), threshold wind characteristics, vegetative cover effects, and anthropogenic effects remain a major key in building explanations of the blowing dust phenomena. If we assume that a seasonal moisture factor adequately represents the effects of several natural processes controlling dust generation, there is still a high degree of unexplained variance in the interannual variability of dust frequency that may relate to large scale disturbance of the desert surface.

References

Brazel, A.J. and R.D. Ziriax 1979. *Monthly and seasonal spatial variability of precipitation: Phoenix and environs.* Climatological Publications, Laboratory of Climatology, Arizona State University, Scientific Paper No. 10.

Brazel, A.J. and W.G. Nickling   In press. The relationship of weather types to dust storm generation in Arizona (1965-1980). *Journal of Climatology.*

Changery, M.J. 1983. *A dust climatology of the western United*

States. Prepared for the Division of Health, Siting and Waste Management Office of the Nuclear Regulatory Commission by the National Climatic Data Center, Asheville, NC.

Gillette, D.A., J. Adams, D. Muhs and R. Kihl 1982. Threshold friction velocities and rupture moduli for crusted desert soils for the input of soil particles into the air. *Journal of Geophysical Research* 87, 9003-15.

Goudie, A.S. 1978. Dust storms and their geomorphological implication. *Journal of Arid Environments* 1, 291-311.

Goudie, A.S. 1983. Dust storms in space and time. *Progress in Physical Geography* 7, 502-30.

Hales, J.E., Jr. 1975. A severe southwest desert thunderstorm: 19 August 1973. *Monthly Weather Review* 105, 1609-13.

Idso, S.B., R.S. Ingram and J.M. Pritchard 1972. An American haboob. *Bulletin of the American Meteorological Society* 53, 930-33.

Idso, S.B., R.J. Reginato and R.D. Jackson 1979. Calculation of evaporation during the three stages of soil drying. *Water Resources Research 15, 487-88.*

Ingram, R.S. 1976. *Summer duststorms in the Phoenix area.* Arizona National Weather Service, Technical Memorandum AZ 1.

Karl, T.R. and R.W. Knight 1985. *Atlas of Palmer Drought Severity Indices (1931-1983) for the contiguous United States.* National Climatic Data Center, Asheville, NC, No. 11.

Karpiscak, M.M. 1980. *Secondary succession of abandoned field vegetation in southern Arizona.* Unpublished   Ph.D. University of Arizona.

Lowe, C.H. 1964. *Arizona landscapes and habitats in vertebrates of Arizona.* Tucson: University of Arizona.

Martin, S.C. and R.M. Turner 1977.   Vegetation change in the Sonoran desert region, Arizona and Sonora. *Journal of the Arizona Academy of Science* 12, 59-69.

Mather, J.R. 1974. *Climatology: fundamentals and applications.* New York: McGraw-Hill.

Middleton, J.N. 1984. Dust storms in Australia: frequency, distribution, and seasonality. *Search* 15, 46-7.

Miller, T.A. In press. Vegetation cover, climate, and Arizona dust. In *Proceedings, Arizona-Nevada   Academy   of Science, Annual Meeting.* Glendale, AZ.

Muller, R.A. 1977.   A synoptic climatology for environmental baseline analysis: New Orleans. *Journal of Applied Meteorology* 16, 20-33.

Nickling, W.G. and A.J. Brazel 1984. Temporal and spatial characteristics of Arizona dust storms (1965-1980). *Journal of Climatology* 4, 645-60.

Orgill, M.M. and G.A. Sehmel 1976. Frequency and diurnal variation of dust storms in the contiguous United States of America. *Atmospheric Environment* 10, 813-25.

Palmer, W.C. 1965. *Meteorological drought.* Research Paper, U.S.

Weather Bureau, U.S. Department of Commerce, No. 45.

Péwé, T.L. 1981. Desert dust: an overview. In *Desert dust: origin, characteristics, and effect on man*, T.L. Péwé (ed.), 1-10. Special Paper Geological Society America, No. 186.

Péwé, T.L., E.A. Péwé, R.H. Péwé, A. Journaux and R.M. Slatt. 1981. Desert dust: characteristics and rates of deposition in Central Arizona, U.S.A. In *Desert dust: origin, characteristics, and effect on man*, T.L. Péwé (ed.),169-89. Special Paper Geological Society of America, No. 186.

Pollard, M.C. 1978. Guidelines for forecasting dust storms in the Southern Great Plains. *National Weather Digest* 3, 40-4.

Sharon, D. 1974. On the modelling of correlation functions for rainfall studies. *Journal of Hydrology* 22, 219-24.

Wigner, K.A. and R.E. Peterson 1985. Duststorms and blowing dust on the Texas South Plains. In *17th conference agricultural and forest meteorology and 7th conference biometeorology and aerobiology*. Preprint volume, 173-7. American Meteorological Society: Scottsdale, Arizona.

15

Maneuver-caused wind erosion impacts, South Central New Mexico

Richard A. Marston

Abstract

Thresholds, rates, and directions of aeolian sediment transport are documented for tank maneuver areas on the Fort Bliss Military Reservation in South Central New Mexico. This portion of the Tularosa Basin is dominated by rounded and elongated coppice dunes fixed by mesquite (*Prosopis glandulosa*) with small, isolated grassland (*Bouteloua eriopoda*) flats. The threshold wind velocity for sand transport, V_t, is a function of antecedent precipitation. Values of V_t at the standard 1 m height range from 3.7-7.2 ms^{-1}. Sand transport is best modeled as a third-order polynomial function of $V-V_t$, with annual rates of transport up to 423 000 gm^{-1}. The effects of tank maneuvers on V_t are minor and short-lived. However, impacts to dune microtopography are more persistent due to preservation of impacts by a surface crust of fines and soluble salts. No sand transport was observed in grassland flats where an algal crust is present.

Introduction

The Dona Ana-Oro Grande Range Complex on the Fort Bliss Military Reservation is used for combat training exercises with heavy tanks and ground-based artillery. The heavy tanks truncate and bisect dunes fixed by mesquite (*Prosopis glandulosa*) and enhance interdune deflation by mechanical break-up of a surface crust of carbonate salts and fine-sized sediment. The magnitude and persistence of maneuver impacts on sand transport and dune microtopography must be evaluated relative to fluctuations in aeolian processes during the Holocene and historical times.

The objective of the research is to determine thresholds, rates, and directions of aeolian sediment transport within maneuver areas subject to variable effects of climate, terrain, and tank activities. Findings will become part of the on-going Mission Environmental Impact Assessment for Fort Bliss. Moreover, the research is designed to provide reliable criteria for scheduling maneuver activities and for designing rangeland rehabilitation projects. On a broader scale, findings will satisfy research needs related to desertification processes as identified by Paylore and Haney (1976) and Sabadell *et al.* (1982).

Literature review

The erosional effects of vehicles in arid regions have received greatest attention in planning studies and environmental impact analyses of off-road vehicles on public lands (e.g. Gillette *et al.* 1980, Lyners *et al.* 1980, Nakata *et al.* 1976, Sheridan 1979, U.S. Bureau of Land Management 1977, 1980, Wilshire 1980). The response of loose, near-surface earth materials to vehicular traffic in arid regions is twofold:

(a) accelerated deflation by wind follows disturbance of the vegetation cover, armor layer of coarse particles, or surface crust of chemically-bound particles, fungal filaments, algae or lichens; and

(b) changes in microtopography due to vehicular traffic are persistent. For example, Sheridan (1979) notes: "some of the tracks made by General Patton's tanks and jeeps in the eastern Mohave during training maneuvers more than 35 years ago are still clearly visible."

Problems caused by sand and dust movement have been outlined by Cook *et al.* (1982) and Goudie (1983) in terms of the responsible aeolian process: deflation-abrasion, transport, and deposition. Deflation-related problems include lost soil fertility (by depleting fines and organics), removal of seeds, exposing plant roots, and scouring and undermining of structures. Sand in transport can cause abrasion of structures and equipment. Dust in transport reduces visibility, resulting in traffic delays, highway accidents, and delaying airline flights. Moreover, dust transport has been associated with dispersal of insects and weeds, suffocation of livestock, interruption of communication services, damage to machinery, and harm to human health. Problems of aeolian deposition include filling of ditches and canals, burial of crops, blocking roads, and contamination of food and drinking water. Ongoing studies will further clarify the significance of dust storms to rainfall acidity, climate change, ocean and stream sedimentation, ground water salinization, and the formation of caliche, duricrusts, loess, and desert varnish (Bomar 1983, Péwé 1981).

Study area

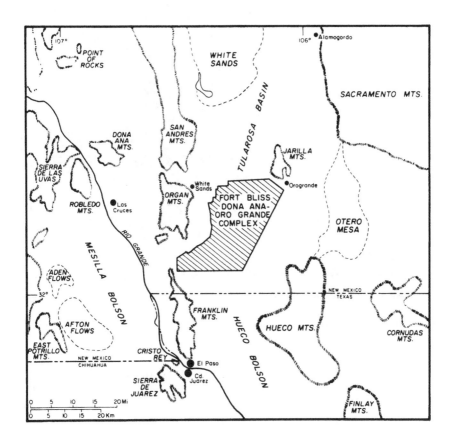

Figure 1 Physiographic map of the study region and location of the Fort Bliss Dona Ana-Oro Grande Complex that served as the study area.

Study area sites were selected on the Dona Ana-Oro Grande Complex, Fort Bliss Military Reservation in South Central New Mexico (Fig. 1). Fort Bliss is the largest military reservation in the United States in terms of area (4503 km²), supported population (93 566), and number of combat soldiers. Since 1944, the Dona Ana-Oro Grande Complex has been used for military training exercises with heavy tanks and ground-based artillery. In 1979, several areas of exceptional archaeological or ecological value were set aside for protection from maneuvers (Fig. 2). Field sites located in these protected areas afford a baseline against which subsequent maneuver impacts in other areas may be assessed.

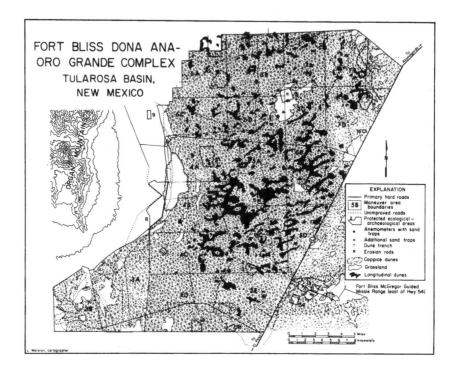

Figure 2 Landtypes and location of study sites in the Fort Bliss Dona Ana-Oro Grande Complex and vicinity.

The Tularosa Basin is located in the Mexican Highland section of the Basin and Range physiographic province (Fenneman 1931, Thornbury 1965). Hawley (1975) designated this portion of the Mexican Highland as the Bolson subsection, characterized by broad, interconnected desert basins, discontinuous mountain ranges, and internal drainage. The Tularosa Basin is a graben complex initiated by crustal tension during the Tertiary by Rio Grande rift activity (Chapin & Seager 1975). The U.T. El Paso layered density-contrast model applied to gravity data for the Tularosa Basin reveals a thickness of up to 2000 m of alluvial and lacustrine sediments. Elevations range from 1220 to 1250 m in Dona Ana-Oro Grande Complex, with a slight regional slope toward the west. Although the Tularosa Basin exhibits internal drainage, inspection of remotely sensed imagery indicates possible relict drainage features aligned along north-south trending Quaternary fault scarps. The Hueco Bolson to the south of the Tularosa Basin exhibits poorly developed drainage toward the Rio Grande. The interconnected Tularosa Basin and Hueco Bolson are bordered by fault block mountains trending north-south. These ranges are the Organ and Franklin mountains on the west, and the Sacramento and Hueco mountains on the east.

The study area is within the Chihuahuan Desert climatic region delineated by Schmidt (1979). This region is characterized by small and variable precipitation, large annual and diurnal temperature changes, low relative humidity, and abundant sunshine throughout the year. Precipitation in the area occurs almost exclusively as rain, with approximately one-half of the average annual total of 210 mm falling in the period July-September. High summer daytime temperatures, low relative humidity, and occasional winds of moderate intensity combine to impose a high average annual potential evapotranspiration of 2030 mm. The large deficit between precipitation and potential evapotranspiration insures persistent soil water shortages, restricting vegetation to species that characterize the Chihuahuan desert scrub communities (Brown 1982). Major wind events in the study region begin with penetration of cold fronts from the north, particularly between February and May when an intense low pressure cell becomes established over the region (Bomar 1983).

Landtypes of the study area

Mapping landtypes as an association of landforms, soils, and vegetation provides a basis for selecting study sites on the Fort Bliss Dona Ana-Oro Grande Complex. Photographic Interpretation Corporation (1979) prepared a set of transparent overlays for photomosaics (1:25 000), including separate maps of landforms, vegetation, drainage, historical and cultural features. A more integrated approach was pursued by Satterwhite and Ehlen (1980) who found that landforms identified from aerial photographs (1:50 000) provide a reliable indicator of soil and vegetation conditions.

Coppice dunes
All past studies identify coppice sand dunes as the major cover type of the study region. Melton (1940) states:

> "... on the disappearance of grasses and other effective sand binders with climatic change, overgrazing, etc., remaining clumps of shrubbery may trap a noticeable amount of blowing sand. Mesquite bush ... grows vigorously on loose sand and is not readily killed by slow sand burial. Sand which falls within the bush may thus stay for a considerable time. If this process continues, a mound of sand eventually is built and held together by a coppice ... Shrub-coppice dunes supported by mesquite bush are present in vast numbers in southeastern New Mexico and in adjoining districts in the southern High Plains. A typical shrub-coppice dune series, barring a reversal of the climatic trend, is the precursor of greater sand concentrations to come; it is

the first topographic manifestation of increasing aridity on a smooth surface of shallow sand, which was previously anchored by grasses and associated vegetation."

The stages in vegetation succession of *Prosopis* are described by (Campbell 1929). Melton (1940) notes that mesquite is not killed by slow burial, but scouring on the windward side will expose roots and kill the plant via sand abrasion. Sand avalanche deposits on the lee side of the dune comprise soil for advancing shrub, usually with large sets of cross-strata dipping at or near the angle of repose (McKee 1982). Coppice dunes grow in height at least as much by interdune deflation as by vertical accretion. To support this notion, Gile *et al.* (1981) cite evidence of smaller hue and percent clay content in the dune and interdune soils.

A feature of the coppice dune landtype salient to considerations of sand transport is a transient surface crust in interdune areas. This crust is composed of fine-sized particles, the cohesion between which is enhanced by elevated concentrations of carbonate salts. The crust is often only a few grains thick, but acts as an armor layer over material below. The composition of the crust was confirmed by physical and chemical comparisons of the crust and particles directly below it. Statistically significant (at $p < 0.01$) higher concentrations of percent silt-clay (6.2% versus 3.6%) and soluble carbonate content (3.7% versus 1.8%) were found in the crust than in the underlying deposits. The following scenario has been responsible for generating a crust in other situations (Chepil & Woodruff 1963). Soluble salts (especially carbonates) precipitate in magnified concentrations at interdune surface when frequent, low intensity rainfall occurs simultaneously with high rates of evaporation. Surface runoff during infrequent, high-intensity rainfall events washes clay and silt into pore spaces of interdune soils. Both types of rainfall events lead to granular binding in interdune areas. Rainfall events of moderate intensity may act to disrupt the surface crust by splash impact and downward leaching of salts.

Inspection of coppice dunes on aerial photographs for the study area and in the field reveals zones of contrasting plan view shape for coppice dunes that has not been distinguished in previous mapping projects (Fig. 2). Coppice dunes are classified as rounded if the ratio of plan view lengths of the longest axis, a, to the perpendicular axis, b, is less than or equal to 1.6. The coppice dunes are classified as elongated if this ratio exceeds 1.6. Rounded coppice dunes occupy approximately 50% of the Dona Ana-Oro Grande Complex with a mean height of 1.5 m. Study site 2 was selected to characterize the rounded coppice dune landtype. The remaining coppice dunes are elongated in a SSW-NNE direction, with a mean height of 2.0 m. Pigott (1978) showed that elongated coppice dunes in the study region are oriented in a direction approximately 45 degrees oblique to the prevailing direction of strongest winds, WSW. Study site 3 was selected to characterize the elongated coppice dune type.

Gile (1981) postulates that the difference in height was controlled by the depth to a buried argillic (Bt) horizon at the time dune formation was initiated. The Bt horizon would act to limit interdune scouring. Greater sand supply in areas of deeper Bt horizons would accumulate in higher dunes. However, abrasion during extreme winds has been observed to scour into the interdune argillic horizon. A horizon of carbonate accumulation could provide the same effect. Satterwhite and Ehlen (1980) found a higher percent of silt-clay as dune size increases, suggesting a possible relationship deserving further investigation.

Soils in the coppice dune areas are sandy loam, sandy clay loam, sandy clay, and silty clay loam. In all cases, the gravel content is less than 3%. A horizon of stage III carbonate accumulation (many nodules and intergranular fillings) is typically found at depths of 40 to 60 cm below the interdune surface, extending horizontally beneath the dunes. From this observation the dunes may be interpreted as having formed long after formation of the petrocalcic horizon. In terms of classification, soils reworked by aeolian processes into coppice dunes are torripsamments (entisols), with no apparent A or B horizon development. The relict horizons that now control depth of scour are paleargids.

Vegetation in the coppice dune areas is homogeneous and dominated by mesquite (*Prosopis glandulosa*) with a ground cover of 6 to 25%. Mesquite roots commonly extend to depths of 15 m. Adventitious roots grow from buried stems at a depth of about 15 cm below the dune surface. Other shrubs include broom snakeweed (*Xanthocephalum sarothrae*), and fourwing saltbush (*Atriplex canescens*), each with ground covers less than 5%. Grass cover ranges from 0 to 25%, consisting mainly of dropseed (*Sporobolus* spp.) and vasey (*Muhlebergia setifolia*) (Satterwhite & Ehlen 1980).

Grassland flats
Small grassland areas on flat terrain occur in isolated locations. Satterwhite and Ehlen (1980) identified five discrete grassland communities in the study area:

(a) *Bouteloua eriopoda*   - *Bouteloua curtipendula*,
(b) *Bouteloua curipendula*   - *Bouteloua uniflora*,
(c) *Scleropogon brevifolius* - *Hilaria mutica*,
(d) *Sporobolus cryptandrus*   - *Sporobolus flexuosus*; and
(e) *Sporobolus giganteus*.

With the exception of the second community listed above, soil depth generally exceeds 30 cm. Shrubs are negligible. Satterwhite and Ehlen (1980) claim that soil depth, soil texture, and soil moisture are the most important environmental factors preserving the boundary between mesquite dunelands and grasslands.

A grassland study site on the western edge of the Fort Bliss McGregor Range was selected to characterize this landtype (Fig. 2). It is dominated by the *Bouteloua eriopoda - Bouteloua curtipendula* community. A surface crust is present at this site, but not as well established as that in the coppice dune landtype and not composed of excess fines or carbonate salts. Instead, bare soil surfaces are covered with cryptogams, the exact nature of which is the subject of ongoing investigations.

Procedures

Two arrays of sand traps were placed at a grassland flats site, a rounded coppice dune site protected from maneuver impacts, a rounded coppice dune site subject to maneuver impacts, and an elongated coppice dune site (Fig. 3). The eight traps in each array face away from the dune they surround in a different compass direction. The type of sand trap selected for the present investigation is that developed by Leatherman (1978), with a trap efficiency of 70% at all particle sizes. The trapped sand was analyzed for particle size distribution and elemental composition. Bivane anemometers were installed at the study sites to relate hourly winds to sand transport and to examine the effects of terrain microrelief on vertical velocity profile. The anemometers were 5.1 m above the ground surface.

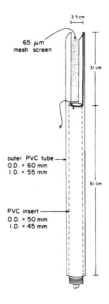

Figure 3 Dimensions of the Leatherman-type sand trap used in the present investigations.

The threshold wind velocity for sand transport, V_t, was determined for each of 40 sampling intervals over a one year period by examining the tabulated frequency data for hourly winds of all velocities (to a resolution of 0.447 ms^{-1}), from eight different compass directions. The direction of maximum wind velocity from which no sand transport was observed gives a lower limit to V_t for that interval. For example, during interval 1 (April 5-12, 1982), no sand was transported from the NE, E, S, SE, and NW. Among those directions, one hour was recorded from the south with an average velocity of 6.3 ms^{-1}. During this interval, sand was collected in traps facing the N, SW, and W; winds exceeded 6.3 ms^{-1} in all three directions. Hence, the threshold wind velocity for this interval was 6.3 ms^{-1}, measured at the height of the anemometer.

Interest in the present investigation centered on two salient controls of V_t: the surface crust described earlier and the impact of heavy tanks and ground-based artillery during training maneuvers. The Army vehicles truncate and bisect dunes and break up the surface crust (Fig. 4). The number of vehicle-days in each maneuver area have been tabulated for each of the 40 intervals. The data represent the degree of training reserved by Army personnel in a given maneuver area, and not necessarily the actual amount or place it later occurred. Because the surface crust in coppice dune areas was observed to be a transient feature dependent on antecedent precipitation and maneuver activity, an attempt was made to model V_t as a function of these factors. Antecedent-precipitation was represented by the maximum 24-hour precipitation (I), and by the number of days per month (normalized from number of days per interval) with precipitation (F) during each interval for which V_t had been determined. Data from the El Paso weather station had to be used, although arid region variability in precipitation renders long-distance extrapolation of data of questionable validity. The vehicle-days for each interval were reduced to the number of "effective vehicle-days" (D), by applying different weighting factors to data from those maneuver areas located upwind and in close proximity to sand traps during the interval. Also, a fraction of the total vehicle-days from previous intervals was included if no precipitation had occurred in the interim.

To relate sand transport to wind velocity, the following calculations were made:

(1) Find $(V_{ij} - V_t)t_{ij}$, where V_{ij} = velocity of ith speed group (increments of 0.447 ms^{-1} in j^{th} direction (one of eight compass directions), and t_{ij} = number of hours wind blew within i^{th} speed group and j^{th} direction;

(2) for each direction, sum the values of $(V_{ij} - V_t)t_{ij}$ to give a force vector,

Figure 4 Impacts to microtopography of rounded coppice dunes casued by heavy tanks.

$$W = \sum_{j=0}^{8} \sum_{i=1}^{n} (V_{ij} - V_t)t_{ij}; \tag{1}$$

(3) for each direction, divide W by the number of hours that $V > V_t$; and

(4) perform regression analyses treating sand transport, q $(gm^{-1}h^{-1})$, as the dependent variable, and the force vector, W, as the independent variable.

Results

Three landtypes were identified in the study area with contrasting implications to aeolian process form: grassland flats, rounded coppice dunes, and elongated coppice dunes. The two dune landtypes contrasted on the basis of microtopography, plan view shape, vertical wind profile, mean and median particle size of trapped sand, sorting of trapped sand, and separate models of sand transport. Grassland flats were distinguished from the dune sites on the basis of microtopography, vertical wind velocity profile, vegetation, and soil depth. The complete lack of aeolian transport in grassland flat sites is the most noteworthy contrast with coppice dune sites.

The following equations were derived to describe threshold wind velocity for the dune landtypes:

$$V_t = 7.85 + (5.35 \times 10^{-2})(I) + (2.69 \times 10^{-4})(I)^2 \qquad (2)$$

$$r^2 = 0.537$$

$$V_t = 7.36 - (8.39 \times 10^{-2})(I) + (2.45 \times 10^{-3})(I)^2 + \qquad (3)$$

$$(0.19)(F) - (5.06 \times 10^{-3})(F)^2 +$$

$$(2.17 \times 10^{-4})(F)^3,$$

$$r^2 = 0.84$$

$$V_t = 7.41 - (8.79 \times 10^{-2})(I) + (2.49 \times 10^{3})(I)^2 + \qquad (4)$$

$$(0.251)(F) - (1.5 \times 10^{-2})(F)^2 +$$

$$(5.88 \times 10^{-4})(F)^3 - (2.40 \times 10^{-6})(D),$$

$$r^2 = 0.89$$

All relationships are significant at $p < 0.01$. These relationships reveal that precipitation intensity provides the greatest explanation power of any one variable. However, the prediction of V_t is much improved by adding precipitation frequency as a second variable (Fig. 5). The r^2 value is surprisingly high considering the spatial disparity between the El Paso weather station where precipitation data were collected, and the study sites, where V_t was measured. The number of vehicle-days provides the least explanation power as a single variable, but does explain nearly one-third of the remaining variance after accounting for precipitation variables. The three-variable model (Eqn 4) supports the hypothesis presented earlier regarding precipitation controls on the surface crust. The model also shows that increasing maneuver activity will decrease the threshold velocity for sand transport. However, over 416 000 effective vehicle-days would be required in any one interval to cause a decrease in V_t of 1.0 ms^{-1}. This value was never exceeded during the study; the maximum number of effective vehicle-days recorded was 5832 during interval 15. The three-variable model also demonstrates that threshold velocities exhibit rapid recovery following heavy maneuver activity, to values reflecting the regeneration of a surface crust with the next precipitation event.

The threshold wind velocity for sand transport, V_t, has been modelled as a function of antecedent precipitation, and to a lesser extent, on maneuver activities. More specifically, V_t depends on the degree of development of a transient surface crust, composed of carbonate salts and fine-sized particles in coppice dune sites, and composed of cryptogams in grassland flat sites. Values of V_t range from 6.3 to 12.1 ms^{-1} as measured by site anemometers, with a mean value of 8.3 ms^{-1}. Using the relationships for drag velocity in each landtype, values measured at the anemometer can be reduced to their equivalent velocity at the standard 1 m height for comparison with other studies. The range in values of V_t above correspond to 3.7-3.8 ms^{-1} for the grassland site, 4.7-4.9 ms^{-1} for the rounded coppice dune site, and 6.8-7.2 ms^{-1} for the elongated coppice dune site. These values generally exceed those for desert dunes reported by Gillette *et al.* (1980). Hourly winds exceeding V_t occurred at a frequency of only 3.9% during the study period, with 70.2% of those winds from the west, 10.2% from the east, and 9.8% from the southwest.

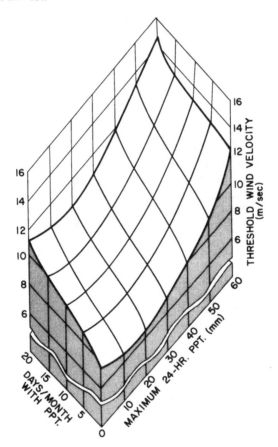

Figure 5 Threshold wind velocity for sand transport as a function of antecedent precipitation.

Rates of sand transport were modelled separately for rounded coppice dunes in protected sites, rounded coppice dunes in sites impacted by maneuvers, and elongated coppice dune sites. This approach to modeling sand transport is similar to the approach of Skidmore (1965) and Skidmore and Woodruff (1968), although they raise $V-V_t$ to the third power. In the present investigation, it was not assumed that sand transport is proportional to the third power of $V-V_t$. In fact, rates of sand transport were best modelled with third-order polynomial equations. For elongated coppice dunes:

$$q = 99.3 + (7.56 \times 10^2)(V-V_t) - \tag{5}$$

$$(1.37 \times 10^2)(V-V_t)^2 + (12.4)(V-V_t)^3 ,$$

$$r^2 = 0.873 ;$$

For rounded coppice dunes in sites protected from maneuver activity:

$$q = 27.1 + (1.61 \times 10^2)(V-V_t) - (36.1)(V-V_t)^2 \tag{6}$$

$$+ (4.4)(V-V_t)^3 ,$$

$$r^2 = 0.912 ;$$

For rounded coppice dunes in sites subject to maneuver impact:

$$q = 39.2 + (1.53 \times 10^2)(V-V_t) - (20.7)(V-V_t)^2 + \tag{7}$$

$$3.7(V - V_t)^3 ,$$

$$r^2 = 0.854 .$$

All relationships were significant at $p < 0.01$; the relationships were also significantly different from each other at $p<0.01$. No sand transport was observed in the grassland site, even during the intervals of strongest winds. This finding is testimony to the effectiveness of grass cover and cryptogamic soil in landscape stability (Van den Ancker et al. 1985). Rates of transport were greatest in the elongated coppice dune site, exceeding by two to four times that recorded in the rounded coppice dune sites. The difference is attributed to greater sand supply in the elongated coppice dune areas. Within the rounded coppice dune area, the rates of transport were greater in the impacted sites than in the protected sites. The curves probably deviate from a third-order power curve at low velocities because the surface crust is protecting loose underlying sand until abrasion breaks up the crust.

Overall, the rates of transport were far below those estimated in models by Bagnold (1941). For instance, when V is 10 ms^{-1} at a height of 5.8 m above the mean aerodynamic surface, Bagnold estimates sand transport at 38 000 $gm^{-1}h^{-1}$.

Sand transport in the elongated dune site using Equation 5 above is estimated at 2260 $gm^{-1}h^{-1}$. The tendency of Bagnold's model to overestimate sand transport was documented in tracer studies by Berg (1983) who noted that Bagnold did not vary V_t in his model as a function of particle and surface conditions. In the present study, the surface crust clearly acted to depress transport rates in the manner described by Gillette *et al.* (1980).

Table 1 Total sand transport by landtype and direction: April, 1982 – March, 1983. Values for each landtype are averages between two arrays at each site.

Dir.	Rounded Dunes-Protected g/m	%	Rounded Dunes-Impacted g/m	%	Elongated Coppice Dunes g/m	%	Total g/m	%
N	1,707	1.2	1,982	0.9	4,307	0.8	7,996	0.9
NE	3,294	2.4	1,274	0.6	15,282	2.8	19,850	2.2
E	6,393	4.6	7,467	3.5	29,930	5.6	43,790	4.9
SE	543	0.4	698	0.3	2,636	0.5	3,877	0.4
S	144	0.1	162	0.1	672	0.1	978	0.1
SW	8,975	6.5	10,030	4.6	45,839	8.6	64,844	7.3
W	116,598	84.6	194,344	89.9	436,838	81.4	747,780	84.0
NW	225	0.2	157	0.1	1,084	0.2	1,466	0.2
Total	137,879	100.0	216,114	100.0	536,588	100.0	890,581	100.0

Regardless of landtype, over 84% of total sand movement was from the west (Table 1). Over 90% of sand transport was from the west and southwest combined. Sand transport from the east was less than 5% and all other directions contribute insignificant amounts. Vector analysis revealed that net transport during the study period was from the west at rates of 114 000 gm^{-1} in the rounded coppice dune-protected sites, 197 000 gm^{-1} in the rounded coppice dune-impacted sites and 423 000 gm^{-1} in the elongated coppice dune sites. Differences in sand transport rates between the rounded coppice dunes and elongated coppice dunes cannot be explained by differences in aerodynamic surface roughness, k, first described by Bagnold (1941). Values of k in the present study were calculated at 0.1 m for grassland site, 0.12 m for the rounded coppice dune site, and 0.13 m for the elongated dune site. Instead, the higher rates for the elongated dune site is attributed to the preservation of helical flow patterns by the parallel, and elongated dune ridges. Video tapes of smoke bomb dye carried by the wind visually confirms the greater efficiency of helical turbulence as an erosional agent in the elongated dune areas. Moreover, the interdune surface crust of salts is more extensive in the rounded coppice dune sites than in the elongated coppice dune sites. As Gillette *et al.* (1980) note: "in locations where water does not collect (and, consequently, salt does not accumulate), evaporation lowers soil moisture ... and neither salt crusting   nor moisture content is present to prevent wind erosion."

The relative amount of sand transport accomplished by winds of various frequency can be examined with a graphical technique developed by Wolman and Miller (1960) (Fig. 6). For each sand transport model developed in the present study, the rate of transport is plotted as a function of V-V_t values. Recall that V_t is dynamic so the absolute values on the abscissa change with changes in V_t. Note the secondary mode in V-V_t values at 8.7 ms^{-1}. The product of frequency and rate of transport at corresponding values of V-V_t yields a curve that represents the relative amounts of geomorphic "work" done on the landscape. In dune landtypes, the greatest amount of work is performed by winds of high magnitude and low frequency during the months of February-May (Fig. 6).

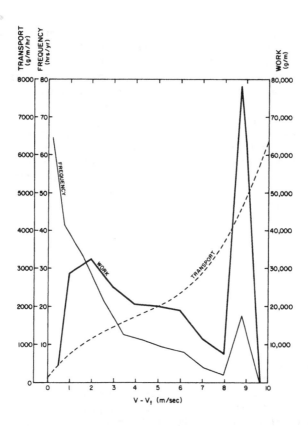

Figure 6 Relative amounts of total sand transport accomplished by winds of various frequency at the elongated coppice dune site.

Training maneuvers with heavy tanks and ground-based artillery affect aeolian process-form in the following ways:

(a) values of V_t are depressed and supply of loose sand are increased for aeolian transport by mechanical break-up of the interdune surface crust, pulverizing particles to finer sizes, as well as truncating and bisecting dunes;

(b) the effects of maneuvers on V_t are minor compared to effects of antecedent precipitation;

(c) the effects of maneuvers on dune microtopography are more persistent due to preservation of the impacts by the same surface crust; and

(d) to minimize impacts, training maneuvers should be scheduled between July and December, when winds are low and V_t is high.

Conclusions

This field-scale study demonstrates that deterministic models of sand transport derived from laboratory data or sites with no microrelief lack application to coppice dune landscapes. Instead, the threshold wind velocity for sand transport is best treated as a dynamic variable in space and time, dependent on the nature and persistence of a surface crust. The contribution of sand from interdune areas is negligible at low values of $V-V_t$ because of the surface crust. At higher values of $V-V_t$, the crust is destroyed to expose fine sand for transport while coarse sand is contributed from the dunes. As a result of these controls, sand transport in the coppice dunes is best modeled as a third-order polynomial function of $V-V_t$ rather than the usual third-order power function. The effect of maneuver traffic on V_t is the basis for analyzing the magnitude and persistence of impacts on sand transport. The effects of maneuvers on threshold wind velocity are short-lived, with values of V_t adjusting quickly upward with precipitation events subsequent to the impact. Additional funding awarded to the author is allowing on-going investigations of the processes that form and destroy the crust under undisturbed and disturbed conditions.

Acknowledgements

This study was funded by the Environmental Office, Directorate of Facilities Engineering, Fort Bliss Military Reservation, in El Paso, Texas. Critical guidance and logistic support for the project was provided by Kevin von Finger, Ecologist with the Fort Bliss Environmental Office. Linda M. Marston drafted the illustrations and Michael Brown typed the manuscript.

References

Bagnold, R.A. 1941. *The physics of blown sand and desert dunes.* London: Methuen.

Berg, N.H. 1983. Field evaluation of some transport models. *Earth Surface Processes and Landforms* 8, 101-14.

Bomar, G.W. 1983. *Texas weather.* Austin: University of Texas Press.

Brown, D.E. (ed.) 1982. Biotic communities of the American southwest: United States and Mexico. *Desert Plants* 4, 1-342.

Campbell, R.S. 1929. Vegetation succession in the Prosopsis sand dunes of southern New Mexico. *Ecology* 10, 392-8.

Chapin, C.M. and W.R. Seager 1975. Evolution of the Rio Grande rift in the Socorro and Las Cruces areas. In *The Las Cruces country: guidebook to the 26th Field conference*, W.R. Seager, R.E. Clemons and J.F. Callender (eds.), 297-322. Socorro: New Mexico Geological Society.

Chepil, W.S. and N.P. Woodruff 1963. The physics of wind erosion and its control. *Agronomy* 15, 211-302.

Cooke, R.U., D. Brunsden, J.C. Doornkamp and D.K.C. Jones 1982. *Urban geomorphology in drylands.* New York: Oxford University Press.

Fenneman, N.M. 1931. *Physiography of the western United States.* New York: McGraw-Hill.

Gile, L.H., J.W. Hawley and R.B. Grossman 1981. *Soils and geomorphology in the Basin and Range area of southern New Mexico: guidebook to the desert project.* Memoir of the New Mexico Bureau of Mines and Mineral Resources, No. 39.

Gillette, D.A., J. Adams, A. Endo, D. Smith and P. Kihl 1980. Threshold velocities for input of soil particles into the air by desert soils. *Journal of Geophysical Research* 85, 5621-30.

Goudie, A. 1983. Dust storms in space and time. *Progress in Physical Geography* 7, 502-30.

Hawley, J.W. 1975. Quaternary history of Dona Ana County region, south central New Mexico. In *The Las Cruces Country: guidebook to the 26th Field conference*, W.R. Seager, R.E. Clemons and J.F. Callender (eds.), 139-50. Socorro: New Mexico Geological Society.

Leatherman, S.P. 1978. A new eolian sand trap design. *Sedimentology* 25, 303-6.

Lyners, M.M., D.L. Weide and E. Von Till Warren 1980. *Impacts: damage to cultural resources in the California desert.* Riverside: U.S. Bureau of Land Management.

McKee, E.D. 1982. *Sedimentary structures in dunes of the Namib desert, Southwest Africa.* Special Paper of the Geological Society of America, No. 188.

Melton, F.A. 1940. A tentative classification of sand dunes: its application to dune history in the southern High Plains. *Journal of Geology* 48, 113-74.

Nakata, J.K., H. Wilshire and G. Barnes 1976. Origin of Mohave Desert dust plumes photographed from space. *Geology* 4, 644-8.

Paylore, P. and E.A. Haney (eds.) 1976. *Desertification: processes, problems, perspectives.* Tucson: University of Arizona Press.

Péwé, T.L. 1981. Desert dust: an overview. In *Desert dust: origin, characteristics, and effect on man*, T.L. Pewe (ed.), 1-10. Special Paper of the Geological Society of America, No. 186.

Photographic Interpretation Corporation 1979. *Fort Bliss environmental analysis and impact assessment.* Report for the Fort Bliss Military Reservation, No. DAAK-70-78-C- 0151.

Pigott, J.D. 1978. Pedology and paleopedology of the western Hueco Bolson for the past two thousand years: a tool for analysis of aboriginal settlement patterns. In *Settlement patterns of the western Hueco Bolson*, M.E. Whalen (ed.), 209-29. Publications in Anthropology, El Paso Centennial Museum, University of Texas at El Paso, No. 6.

Sabadell, J.E., E.M. Risley, H.T. Jorgenson and B.S. Thornton 1982. *Desertification in the United States: status and issues.* Washington, D.C.: U.S. Bureau of Land Management.

Satterwhite, M.B. and J. Ehlen 1980. *Vegetation and terrain relationships in south central New Mexico and western Texas.* Report of the U.S. Army Corps of Engineers Topographic Laboratories, No. ETL-0245.

Schmidt, R.H., Jr. 1979. A climatic delineation of the "real" Chihuahuan Desert. *Journal of Arid Environments* 2, 243-50.

Sheridan, D. 1979. *Off-road vehicles on public land.* Washington, D.C.: U.S. Council on Environmental Quality.

Skidmore, E.L. 1965. Assessing wind erosion forces: direction and relative magnitude. *Proceedings of the Soil Science Society of America* 29, 587-91.

Skidmore, E.L. and N.P. Woodruff 1968. Wind erosion forces in the U.S. and their use in predicting soil loss. *USDA Agricultural Handbook* 346.

Thornbury, W.D. 1965. *Regional geomorphology of the United States.* New York: Wiley.

U.S. Bureau of Land Management 1977. *California desert plan program: planning area analysis.* Riverside: U.S. Department of the Interior.

U.S. Bureau of Land Management 1980. *The California desert conservation area plan.* Sacramento: U.S. Department of the Interior.

Van den Ancker,J.A.M., P.D. Jungerius and L.R. Mur 1985. Short communications: the role of algae in the stabilization of coastal dune blowouts. *Earth Surface Processes and Landforms* 10, 189-92.

Wilshire, H.G. 1980. Human causes of accelerated wind erosion in California's deserts. In *Thresholds in Geomorphology*, D.R. Coates and J.D. Vitek (eds.), 415-533. London: George Allen & Unwin.

Wolman, M.G. and J.P. Miller 1960. Magnitude and frequency of forces in geomorphic processes. *Journal of Geology* 68, 54-74.

16

Rock varnish as an indicator of aeolian environmental change

Ronald I. Dorn

Abstract

Micromorphological changes in rock varnishes indicate long-term fluctuations in the abundance of aeolian dust. Varnishes on K-Ar dated volcanics from the Coso Range, eastern California, record three lengthy dusty periods alternating with two lengthy periods of less-abundant dust over the last ca. 200 000 years BP. This sequence corresponds with the paleo-lake levels of nearby Searles Lake. The presence or absence of micromorphological changes may be used to map the distribution of dusty aeolian conditions in the late-Pleistocene.

Cation-ratio dating of rock varnishes formed on ventifacts provides a minimum age for the cessation of aeolian abrasion. Varnish on ventifacts started to form about 5300 years BP in the Cronise area of the Mojave Desert in eastern California. Varnish on gravel float on a fossil dune in the same area started to form ca. 5100 years BP. These chronometric ages suggest that a major period of aeolian activity abated by the mid-Holocene in the Cronise area of the central Mojave Desert.

Although problems in accurate interpretation of rock varnish still exist, these case studies illustrate the promise of rock varnish as an indicator of aeolian environmental change.

Introduction

There are few terrestrial records of change in aeolian conditions and fewer still that are amenable to dating. In this paper, two new methods of assessing the spatial and temporal variations in aeolian conditions are presented, both using rock varnish. Although still experimental, these methods have the potential to enhance our understanding of aeolian environmental change during the Holocene and late-Pleistocene.

Rock varnish is a paper thin accretion of clay minerals, manganese and iron oxides and trace elements that form dark coatings on rocks (Potter & Rossman 1977). Ubiquitous in the terrestrial weathering environment, varnish occurs from hyper-arid to humid areas (Dorn & Oberlander 1982). While varnishes develop in soils, regolith, riverine, lacustrine, crenitic, cave and other near surface environments, this paper will focus on those varnishes exposed only to subaerial conditions. This is to assure that the constituents of varnish are derived exclusively from airborne fallout.

The nature of airborne fallout affects several characteristics of subaerial varnishes. The manganese:iron ratio in varnish layers may be used to record changes in the alkalinity of the aeolian environment (Dorn 1984). Variations in the stable carbon isotope composition of organic matter in varnish may indicate paleobotanical fluctuations (Dorn & DeNiro 1985).

This paper proposes that the micromorphology of rock varnish is controlled largely by the abundance of aeolian dust, and that superposition of dissimilar micromorphologies reflects fluctuations in the aeolian environment. In addition, cation-ratio dating of varnish (Dorn 1983, Dorn *et al.* 1986) on ventifacted surfaces provides the opportunity to determine a minimum age for the cessation of aeolian abrasion.

Environment and varnish micromorphology

Interest in the micron-scale structure of rock varnish started with scanning electron microscope (SEM) observations by Krumbein (1969). Potter and Rossman (1977), Perry (1979), Dorn and Oberlander (1982) and Whalley (1983) followed suit, but a systematic set of observations on the factors influencing varnish micromorphology has been lacking until now.

Method
Varnishes were sampled for this study from over 100 arid, semi-arid and humid sites, and along over a dozen climatic-altitude transects in western North America. Unless stated otherwise, samples were collected from outcrops at least one meter above the surrounding soil, cleaned with compressed air and light agitation in deionized water to remove loose airborne fallout, and coated with gold-paladium, gold, or carbon for viewing by scanning electron microscopy (SEM) and by energy-dispersive analysis of X-rays (EDAX). Samples were not scrubbed clean, as some investigators have done, because harsh mechanical action alters some varnish structures. Before coating, selected samples were also prepared by critical point drying to examine micro-organisms.

Classification of varnish micromorphologies

While only a selected data set can be presented here, over 1000 hours of SEM observations indicate varnish micromorphologies fall into a continuum of forms that vary between botryoidal and lamellate (Fig. 1). Virtually all botryoidal sites have a vegetation cover of over 40 to 50%. Lamellate and intermediate varnishes tend to occur where the vegetation cover ranges from <5% to ca. 35%.

Subaerial varnishes were also sampled from over a dozen altitude-climatic gradients in western North America. A representative result, from Kitt Peak in southern Arizona, is presented in Figures 2A-2C. The environmental transition between intermediate and botryoidal in these transects is a plant cover of about 40%. No statistically significant relationship was found at these transects between varnish micromorphology and surficial soil texture, lithology, plant height or plant species.

Many more micromorphological structures were observed than can be presented in this paper (Dorn 1985). However, approximately 75% of all subaerial varnishes fall into the botryoidal-lamellate continuum.

Aeolian control on varnish micromorphology

Vegetation cover provides the best correlation between a characteristic of the environment and varnish micromorphology. In presenting a working model of the development of varnish micromorphology, I will endeavor to show that lamellate varnish reflects the abundance of dust in a sparsely vegetated environment, and botryoidal varnish reflects the relative paucity of dust in a well-vegetated area.

Varnish accretion is a two-step process. Initially, micro-organisms oxidize ambient manganese and iron (Krumbein & Jens 1981, Dorn & Oberlander 1982, Taylor-George *et al.* 1983). Manganese and iron are then incorporated into mixed-layer illite-montmorillonite clay minerals that settle onto varnish surfaces (Potter & Rossman 1977, Fig. 2D). The limiting factors on varnish growth are the abundance of airborne clay minerals and the rate of microbial oxidation.

In the model proposed first by Dorn and Oberlander (1982) and supported here, when clay minerals are abundant in the aeolian environment, the deposition of clay platelets in a parallel orientation dominates the varnish structure and a lamellate varnish is produced. When aerosolic clay minerals are less common, oxides accumulate around discrete microbial nuclei and a botryoidal micromorphology is favored.

A test of this model would be a study on the rates of airborne fallout at varnish sites. Airborne collectors were placed for this purpose, and available data indicate that lamellate varnishes occur where dust fallout is greatest. The literature supports these preliminary results; for example, Orgill and Sehmel (1976) and Cahill

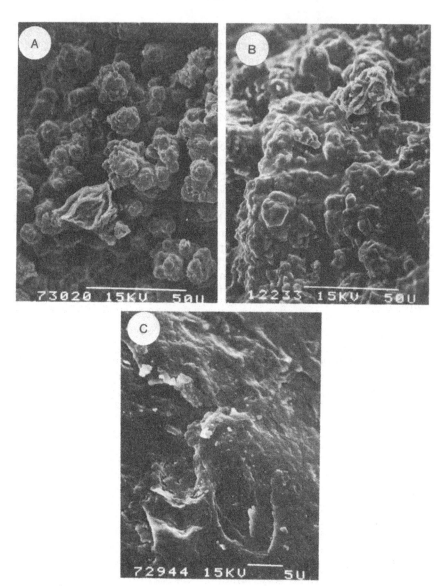

Figure 1 Micromorphology Continuum. Scale bars and scale lengths are in the lower right corner for all scanning electron micrographs in this paper. **A** Botryoidal micromorphology agglomerations of growths of different sizes, collected about 8 km east of Superior, in Devil's Canyon, Arizona. **B** Semi-botryoidal micromorphology, from Superstition Mountains, Arizona. **C** Lamellate varnish created by the parallel deposition of clays and oxides, collected from South Mountain Park, Phoenix, Arizona.

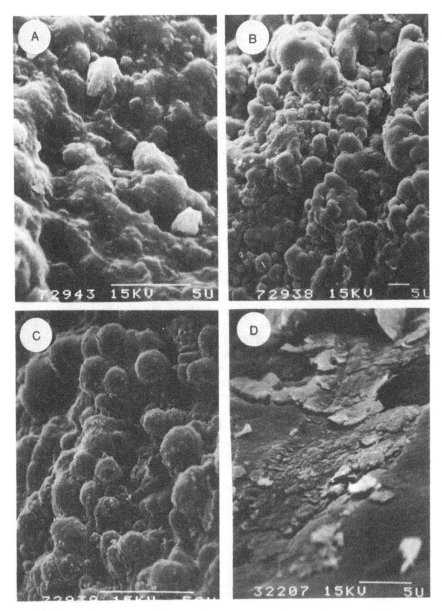

Figure 2 Kitt Peak transect and incipient varnish. Micromorphologies of varnishes sampled along an altitude-climatic gradient from the base to just below the top of Kitt Peak, southern Arizona. A Semi-lamellate from 914m. B Between botryoidal and lamellate from 1219m. C Compound botryoidal from 1890m. D Incipient varnish on a young surface of Badwater alluvial fan, Death Valley California.

et al. (1981) indicate that suspended aerosols are more abundant in desert areas of sparse vegetation. Gillette and Walker (1977) and Post and Buseck (1984) found that the aerosol clays in desert areas are often composed of mixed-layer clays, similar to those in rock varnish. Marshall (1971) presented a likely mechanism for the relative lack of dust in well vegetated areas. Where the vegetation cover is dense, it shields the ground surface and increases the threshold velocity required to deflate particles, and Gillette *et al.* (1980) demonstrated that the characteristics of desert surfaces and desert meteorology make arid regions particularly sensitive to the suspension of dust. However, with the very slow rates of varnish formation in arid environments (e.g. Elvidge 1979, Whalley 1983, Dorn *et al.* 1986), contemporary measurements of human-affected airborne fallout may not reflect the aeolian environment that influenced the development of varnish structures.

There are more specific lines of evidence that support this model. First, SEM/EDAX observations indicate that botryoidal varnishes are deposited at discrete loci. Second, in laboratory replications of varnishes (e.g. Dorn & Oberlander 1982), a botryoidal form of development is favored where the media contains a lower concentration of bentonite and illite clay; importantly, the mineralogy of the manganese deposit in the laboratory varnish is birnessite (G. Rossman, personal communication 1983), the same manganese mineral found in the true desert varnish (Potter & Rossman 1977). Lastly, electron microprobe, particle induced X-ray emission (PIXE) and EDAX measurements of the surface layer of varnishes from over 50 sites in the western U.S. indicate that botryoidal varnishes often have significantly greater concentrations of Mn and Fe oxides than intermediate and lamellate varnishes, on the order of 1.2x to over 2x more. This is what would be expected if lamellate varnish represents the overwhelming of botryoids by the deposition of clay minerals.

Discussion

These observations suggest that the relative amount of aeolian fallout controls the micromorphology of varnish. There are, however, certain local conditions where the empirical correlation between the dustiness of the environment and the micromorphology of varnish is disrupted. These potentially confounding factors will be presented so that other researchers will be able to control for these effects in future sampling.

Aeolian abrasion can either completely remove a varnish or greatly disturb its appearance (Dorn & Oberlander 1982). Similarly, lichens, cyanobacteria (Krumbein & Jens 1981), and microcolonial fungi (Taylor-George *et al.* 1983) have all been observed in this study as active agents of biochemical erosion, probably through the secretion of organic acids (e.g. Fig. 3A). At other times, these adventitious organisms simply occur on varnishes, but do not erode them (Fig. 3B). Microcolonial fungi may mimic the appearance of varnish, especially when they are covered with a thin coating of clays. However, no evidence gathered in this study indicates that these organisms play an active role in the concentration of manganese or iron in varnish.

Varnish bacteria can counter the influence of aeolian dust and generate a semi-botryoidal appearance even in arid regions. This occurs in hollows that are microns to millimeters in diameter. These sheltered depressions provide a suitable micro-environment for bacteria to colonize and concentrate Mn-Fe oxides fast enough to generate a micromorphology that trends toward botryoidal.

The structure of a thin, youthful varnish can often be influenced by an irregular shape of the underlying rock. The depositional tendency toward either lamellate or botryoidal is obscured until the varnish thickens to the point where the influence of the substrate is effectively erased.

The correlation between micromorphology and the aeolian environment is invalid where sampling micro-environments are not subaerial and are not well above the surrounding soil. For example, Figure 4A illustrates the boundary of crack varnish and subaerial varnish. The subaerial varnish is botryoidal in tendency, but the clay fallout that has been collecting in the rock crevice probably overwhelmed the subaerial tendency towards botryoidal and generated a lamellate varnish. Figures 4B and 4C contrast varnishes collected over one meter above the soil surface with a varnish collected just above the soil surface. Proximity to the soil surface tends to push the varnish towards the lamellate end of the continuum.

Lastly, there are pseudo-varnishes that may be mistaken for true varnishes and can have a roughly similar micromorphology. For example, Mn-rich hydrothermal deposits often resemble botryoidal varnishes superficially. Also, shiny and smooth accumulations of amorphous silica can occur intermixed with rock varnish (Butzer *et al.* 1979) or as separate coatings (e.g. Curtiss *et al.* 1985).

Where varnish is not influenced by these local effects, there appears to be a strong relationship between varnish micromorphology and environment. Part of the frustration in proving this conclusively is the slow rate of varnish growth. Elvidge (1979) and Whalley (1983) stress that varnish takes thousands of years to completely coat a rock. For example, chert artifacts in the Mojave Desert that have been cation-ration dated by Dorn *et al.* (1986) to be older than 10 000 years BP still do not have a complete varnish coating. Similarly, cobbles on the 10 500 BP high stand of Searles Lake (Smith 1984) are not completely varnished. This suggests that observations of the contemporary environment over a few years may not necessarily relate to the longer-time required to develop the character of surficial varnish structures.

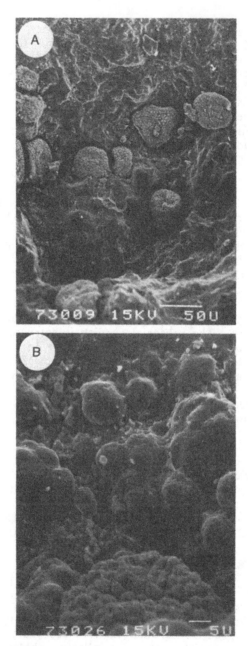

Figure 3 Effect of microcolonial fungi. A MCF creating hollows probably by the secretion of organic acids. Varnish collected near Jake's Corner in central Arizona. **B** Macrobotryoidal varnish (upper part of the image) is clearly distinct from the MCF (lower part of image). Varnish collected near dam of Roosevelt Lake, Arizona.

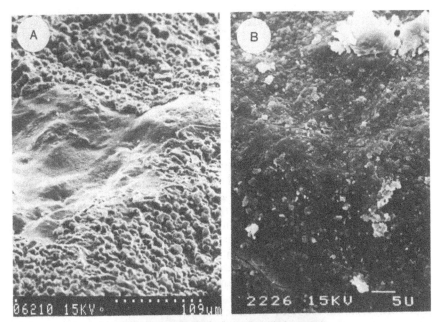

Figure 4 Microenvironmental effects. A Juxtaposition of subaerial varnish (semi-botryoidal) and varnish developed only in a rock crevice (lamellate), collected from a surface of Hanaupah Canyon alluvial fan, Death Valley. **B** Lamellate varnish collected just above the soil surface, from Ah-shish-la-pah badlands, near Chaco Canyon, New Mexico. **C** Microbotryoidal varnish collected a little over a meter above the adjacent soil, from the same site as Figure 4B in an area of Pinyon Pine and dense bunch grass.

Paleo-aeolian record in varnish

Micromorphological stratigraphies with age-control
This section presents microstratigraphies of alternating lamellate and botryoidal layers in varnish on dated surfaces. Accepting even on a tentative basis the validity of the correlation between aeolian environment and varnish micromorphology, fossil layers in varnishes are interpreted here as paleoenvironmental signals, where fossil lamellate layers indicate dusty conditions, and fossil botryoids indicate less aeolian fallout.

Figure 5 presents a collection of micromorphological stratigraphies of varnishes on various volcanic rocks in the Coso field and vicinity (Duffield & Bacon 1981). Each micrograph in Figure 5 is representative of multiple samples taken from adjacent rocks and adjacent outcrops in the same K-Ar dated volcanic deposit. However, the large age-uncertainties in these youthful K-Ar dates should be kept in mind.

McFadden *et al.* (1984) and Mayer and Anderson (1984) suggest a mechanism, operating with plant cover, that may explain dust fluctuations in the Coso Range. As the pluvial lakes in eastern California expanded and contracted, the amount of dust generated decreased and increased concomitantly. The Coso Range is surrounded by Owens, China, and Searles lake basins (e.g. Smith 1984). The lake-level history of Searles Lake over the last 250 000 years BP corresponds approximately with the observed micromorphological changes in varnish (Smith 1984). A dry to shallow lake occurred in the Searles basin from ca. 0 to 10 500 years BP, along with xeric vegetation in the northern Mojave Desert (Spaulding *et al.* 1983, Wells 1983). This may correspond with the top lamellate layer in Coso Range varnishes (e.g. Fig. 5A). The single subsurface botryoidal layer on varnishes younger than ca. 81 000 years BP may correlate with the deep and intermediate lake levels from 10 500 to 90 000 years BP (Figs 5B & C). The second lamellate layer in varnishes older than ca. 81 000 years BP may be from the shallow to intermediate lake phase from about 90 000 to 105 000 years BP (Figs 5D & E). The second botryoidal layer in Figure 5F may have formed during the phase of a deeper lake from about 105 000 to 130 000 years BP. Extensive lamellate deposits occur below the second botryoidal layer in Coso varnishes less than 600 000 years BP; this lamellate varnish may be from the shallow lake from ca. 130 000 to 570 000 years BP. Varnishes on volcanics older than ca. 600 000 to 1 million years BP may have many layers.

There is abundant support for a widespread micromorphological change in areas that have experienced major paleohydrological (Smith 1984, Begin 1984) and paleobotanical fluctuations (Spaulding *et al.* 1983, Wells 1983) in the late-Wisconsin. A problem in interpretation arises when varnishes older than the late-Wisconsin are examined. The major difficulty revolves around whether these

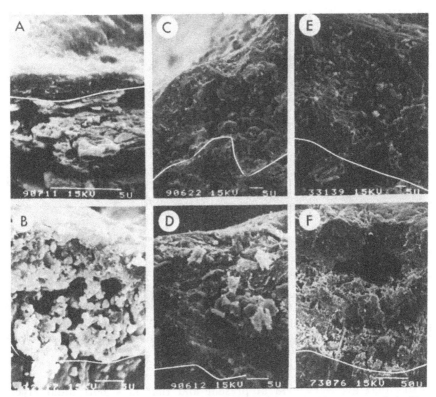

Figure 5 Coso Range micromorphological stratigraphies. Representative varnishes on progressively older surfaces in the Coso Range and vicinity, eastern California. Numbers and K-Ar ages are from Duffield and Bacon (1981). Lines in the images indicate the contact between the varnish and the underlying rock. **A** Entirely lamellate varnish on cobble of ca. 10 500 years BP shoreline of Searles Lake (Smith 1984). **B** Varnish on alluvial fan, west of Coso Range, with a cation-ration date of 26 500 ± 5500 years BP. Sequence of lamellate on botryoidal. **C** Varnish on Coso No. 17 (57 000 ± 16 000 years BP). Sequence of lamellate on botryoidal. **D** Varnish on Coso No. 1 (81 000 ± 8000 years BP). Sequence of lamellate on botryoidal on lamellate. **E** Varnish on Coso No. 28 (140 000 ± 89 000 years BP). Sequence of lamellate on botryoidal on lamellate on indistinct. Note the K-Ar large age-uncertainty. **F** Varnish on Coso No. 26 (234 000 ± 22 000 years BP). Sequence of lamellate on botryoidal on lamellate on botryoidal on lamellate.

layers can be correlated with other records of Quaternary environmental change, or whether they represent only local fluctuations. Until the precision of dating varnishes can be improved for times beyond the period accessible to accelerator-radiocarbon dating (Dorn *et al.* 1986), interpretation of older varnish micromorphological stratigraphies will remain speculative.

Mapping regions of abundant dust in the late-Wisconsin
There are rock varnishes that are at least late-Wisconsin in age at

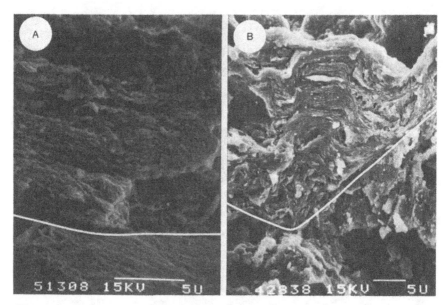

Figure 6 Late-Wisconsin lamellate structures. A Varnish on 534 m shoreline of Manix Lake, accelerator-radiocarbon dated at 16 800 ± 700 years BP (Dorn *et al.* 1986). B Varnish on desert pavement near Bristol Dry Lake, Mojave Desert, cation-ration dated at 25 000 ± 3000 years BP.

sites that are very arid at present and that lack botryoidal layers in the subsurface.   The varnishes in Figures 6A and 6B were collected in the central Mojave Desert well below 1000 m. Spaulding (1983) notes that desert scrub vegetation did occur below 1000 m during the late-Wisconsin in the Mojave Desert. Assuming accuracy in dating, it is possible that these environments were dusty enough even in the late-Wisconsin to maintain a lamellate micromorphology.

It would be possible to use the presence or absence of botryoidal varnishes to provide a means of mapping the distribution of arid areas experiencing significant levels of dust deposition during the late-Wisconsin.

Dating cessation of aeolian abrasion
East Cronise basin in the central Mojave Desert of eastern California contains abundant ventifacts.   Most of these, however, are fossil forms in that varnish can be found growing on the polished, pitted and grooved surfaces.   This varnish was collected and analyzed for cation-ratio dating.

Varnish was sampled from a boulder approximately two meters in diameter on top of a small hill, just southeast of Basin Road exit off I-15, between Barstow and Baker.   The varnish formed on the grooves a few centimeters deep dates to 5500 ± 200 years BP. The varnish formed on grooves a few millimeters deep dates to 5300 ± 300 years BP, essentially identical to the age of the varnish in the deeper grooves. Similarly, ventifacted boulders sampled from the crest of the Cronise Mountains have varnish growing on etched surfaces that cation-ratio dates to 5500 ± 700 years BP.

On the east side of the Cronise Mountains in East Cronise basin, there is a dune known locally as "Cat Dune." The dune is largely fossilized, and the upper layer of sand is indurated. A gravel float has formed a lag on the bottom part of this fossilized sand dune. Cation-ratio dating of varnish on this gravel yields a chronometric age of 5100 ± 200 years BP.

Discussion

The interpretation of paleo-aeolian fluctuations from rock varnish has been greatly simplified. In this section, I will briefly illustrate some of the potential complications involved.

The sample preparation and collection procedures are subjective. Multiple samples are, therefore, needed to ensure that enough exposures are present to verify the reproducibility of the results, from place to place on a single rock, from rock to rock on a single outcrop, and from outcrop to outcrop on the same surface.

Crack varnishes can develop micromorphologies different from the adjacent subaerial varnish (Fig. 4A). When crack varnishes are exposed by spalling, they may preserve the micromorphology from the environment of the crevice and generate a stratigraphy unrelated to changes in the aeolian environment (Fig. 7A).

The thickness of a given varnish micromorphology layer cannot be related to the length of time under a given paleoenvironmental condition. The varnish is simply recording a sequence of paleo-aeolian periods of sufficient intensity to cross a threshold between botryoidal and lamellate micromorphologies. Information on duration is not provided.

It may be possible that a positive feedback process prolongs a given micromorphology, even after an environmental change has occurred. For example, about three quarters of the varnish in a ca. 14 600 year old sample from the Cima volcanic field was found to be botryoidal; however, about two thirds of the time, the varnish has been exposed to a relatively drier Holocene environment (Spaulding *et al.* 1983, Ore & Warren 1971). Related issues are the variable lag effects between a climatic change and the vegetation and lacustrine changes in an area, and the ca. 100 year lag between the exposure of a surface and the onset of varnishing (Dorn *et al.* 1986).

A more serious interpretive problem is the preservation of varnish micromorphologies. While many botryoidal micromorphologies are preserved after burial, not all are. Some botryoidal forms undergo a diagenesis to lamellate structures (e.g.

Fig. 7B). Where subsurface botryoidal layers do occur, some may be laterally discontinuous, gradually passing into entirely lamellate cross-sections, and then passing back into the same sequence of micromorphologies. Extensive EDAX analyses indicate the fossil botryoidal layers that are preserved have a much higher concentration of manganese and iron. Unless a fossil botryoidal layer is indurated with high concentrations of Mn- and Fe-oxides, it can undergo diagenesis into lamellate structures.

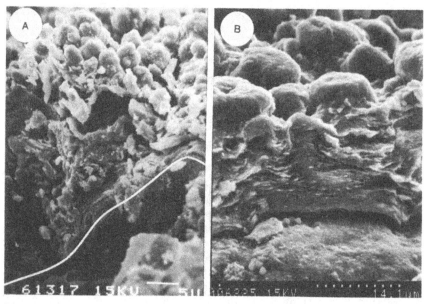

Figure 7 Complications in interpretation. A Botryoidal micromorphology on lamellate, collected from a recently spalled surface, Mid-Hills, Mojave Desert. The lamellate varnish found under the botryoidal layer probably formed while in a rock crevice (e.g. Figure 4A). With spalling and exposure to subaerial processes, a surface botryoidal varnish formed. Therefore, superposition of dissimilar micromorphologies can occur without a major environmental change. B Semi-botryoidal surface micromorphology that is lamellate in cross-section. Sample collected from below the "Blackwelder" shoreline of Lake Manly on Shoreline Butte in Death Valley (Hooke, 1972).

Conclusion

Changes in the micromorphology of subaerial varnishes offer an intriguing potential for recording fluctuations in the level of aeolian dust fallout over the last ca. 200 000 years BP. Varnish micromorphology appears to be an accurate indicator of a change from a less-dusty environment during the late-Wisconsin to a dusty environment in the Holocene. Because varnish is ubiquitous and has good internal age-control, it would be possible to map the areas in the southwest that experienced fairly dry, dusty conditions even during the late-Wisconsin.

Cation-ratio dating of varnishes formed on fossil ventifacts in the central Mojave Desert indicate that major aeolian abrasion ended by about 5100 years BP. These chronometric dates provide the first ages on the cessation of ventifaction, and they correspond with the cation-ratio ages of varnishes formed on rock float on a fossilized dune in the region. Although only a few ages have been obtained, they provide an indication of the potential of varnish for assessing paleo-aeolian activity.

Acknowledgements

Funded partly by NSF graduate fellowship research stipend, National Geographic Grant 2961-84, and NSF grant SES80-24555 to T.M. Oberlander. Thanks to V. Tchakerian for assistance in the collection and analysis of varnish on ventifacts, D. Quimby for help in field collections, C.R. Berger and A.R. Orme for a critical eye on an earlier draft, and T.M. Oberlander for many conversations on varnish.

References

Begin, Z.B. 1984. Lake Lisan levels in the last 30 000 years. *Israel Geological Society Annual Meeting* 17-18.

Butzer, K.W., G.J. Fock, L. Scott and R. Stuckenrath 1979. Dating and context of rock engravings in South Africa. *Science* 203, 1201-14.

Cahill, T.A., L.L. Ashbaugh, R.A. Eldred, P.J. Feeney, B.H. Kusko and R.G. Flocchini 1981. Comparisons between size-segregated resuspended soil samples and ambient aerosols in the western United States. In *Atmospheric aerosols: source/air quality relationships*, E.S. Macias and P.K. Hopke (eds.), 269-85. Washington, D.C.: American Chemical Society.

Curtiss, B., J.B. Adams and M.S. Ghiorso 1985. Origin, development and chemistry of silica-alumina rock coatings from semi-arid regions of the island of Hawaii. *Geochemica et Cosmochimica Acta* 49, 49-56.

Dorn, R.I. 1983. Cation-ratio dating: A new rock varnish age-determination technique. *Quaternary Research* 20, 49-73.

Dorn, R.I. 1984. Speculations on the cause and implications of rock varnish microchemical laminations. *Nature* 310, 767-70.

Dorn, R.I. 1985. *Rock varnish: a key to the Quaternary period in western North American deserts*. Ph.D. Dissertation, University of

California, Los Angeles.

Dorn, R.I. and M.J. DeNiro 1985. Stable carbon isotope ratios of rock varnish organic matter: a new paleoenvironmental indicator. *Science* 227, 1472-4.

Dorn, R.I. and T.M. Oberlander 1982. Rock varnish. *Progress in Physical Geography* 6, 317-67.

Dorn, R.I., D.B. Bamforth, T.A. Cahill, J.C. Dohrenwend, B.D. Turrin, D.J. Donahue, A.J.T. Jull, A. Long, M.E. Macko, E.B. Weil, D.S. Whitley and T.H. Zabel 1986. Cation-ratio and accelerator-radiocarbon dating of rock varnish on archaeological artifacts and landforms in the Mojave Desert, eastern California. *Science* 213, 830-33.

Duffield, W.A. and C.R. Bacon 1981. Geological map of the Coso volcanic field and adjacent areas, Inyo County, California. *U.S. Geological Survey Miscellaneous Investigations* Map I-1200.

Elvidge, C.D. 1979. *Distribution and formation of desert varnish in Arizona*. M.S. Thesis, Arizona State University, Tempe.

Gillette, D.A. and T.R. Walker 1977. Characteristics of airborne particles produced by wind erosion of sandy soil, high plains of west Texas. *Soil Science* 123, 97-110.

Gillette, D.A., J. Adams, A. Endo and D. Smith 1980. Threshold velocities for input of soil particles into the air by desert soils. *Journal of Geophysical Research* 85 (C10), 5621-30.

Hooke, R.L. 1972. Geomorphic evidence for late-Wisconsin and Holocene tectonic deformation, Death Valley, California. *Geological Society of America Bulletin* 31, 2075-98.

Krumbein, W.E. 1969. Uber den Einfluss der Mikroflora auf die Exogene Dynamik (Verwitterung unde Krustenbildung). *Geologische Rundschau* 58, 333-63.

Krumbein, W.E. and K. Jens 1981. Biogenic rock varnishes of the Negev Desert (Israel): an ecological study of iron and manganese transformation by cyanobacteria and fungi. *Oecologia* 50, 25-38.

Marshall, J.K. 1971. Drag measurements in roughness arrays of varying density and distribution. *Agricultural Meteorology* 8, 269-92.

Mayer, L. and M. Anderson 1984. Variations in carbonate dust flux in the southwestern United States--implications for rates of development of carbonate soil horizons. *Geological Society of America Abstracts with Programs* 16, 586-87.

McFadden, L.D., S. Wells, J.C. Dohrenwend and B. Turrin 1984. Cumulic soil formation in eolian parent materials on flows of the Cima volcanic field, Mojave Desert, California. In *Surficial geology of the Eastern Mojave Desert, California*, J.C. Dohrenwend (ed.), 134-49. Reno Meeting: Geological Society of America.

Ore, H.T. and C.N. Warren 1971. Late Pleistocene-early Holocene geomorphic history of lake Mojave, California. *Geological Society of America Bulletin* 82, 2553-62.

Orgill, M.M. and G.A. Sehmel 1976. Frequency and diurnal variations of dust storms in the contiguous USA. *Atmospheric Environment* 10, 813-25.

Perry, R.S. 1979. *Chemistry and structure of desert varnish.* M.Sc. Thesis, University of Washington, Seattle.

Post, J.E. and P. Buseck 1984. Characterization of individual particles in the Phoenix urban aerosol using electron-beam instruments. *Environmental Science and Technology* 18, 35-42.

Potter, R.M. and G.R. Rossman 1977. Desert varnish: importance of clay minerals. *Science* 196: 1446-48.

Smith, G.I. 1984. Paleohydrologic regimes in the southwestern Great Basin, 0-3.2 myr ago, compared to other long records of "global" climate. *Quaternary Research* 22, 1-17.

Spaulding, W.G. 1983. Late-Wisconsin macrofossil records of desert vegetation in the American southwest. *Quaternary Research* 19, 256-64.

Spaulding, W.G., E.B. Leopold and T.R. Van Devender 1983. Late Wisconsin paleoecology of the American Southwest. In *Late-Quaternary Environment of the United States* Volume 1. *The Late Pleistocene*, S.C. Porter (ed.), 259-93. Minneapolis: University of Minnesota.

Taylor-George, S., F. Palmer, J.T. Staley, D.J. Borns, B. Curtiss and J.B. Adams 1983. Fungi and bacteria involved in desert varnish formation. *Microbial Ecology* 9, 227-45.

Wells, P.V. 1983. Paleobiogeography of montane islands in the Great Basin since the last Glaciopluvial. *Ecological Monographs* 53, 341-82.

Whalley, W.B. 1983. Desert varnish. In *Chemical sediments and geomorphology*, A.S. Goudie and K. Pye (eds.), 197-226. London: Academic Press.

Index